Praise for
Essential Green Roof Construction

In this impressive book, Leslie Doyle shares her wealth of experience and knowledge installing and maintaining green roofs. The writing is detailed and comprehensive and, importantly, easy for the average person to understand. Filled with photos, real-life case studies, and practical guidance, *Essential Green Roof Construction* left me with the confidence to get started with a green roof. I have some clear steps and ideas to move forward and feel like I know the right questions to ask.

—Stephen Hill, associate professor,
School of the Environment, Trent University

For a thorough understanding of what constitutes a modern green or living roof, this essential guide will provide all you need to know to install and maintain a beautiful and ecological compliment to your home or office. The details are clearly described and illustrated and safety and durability are emphasized throughout, while specific engineering analysis is left for professionals on a case by case basis.

—Kelly Hart, GreenHomeBuilding.com, author
Essential Earthbag Construction

Required reading for anyone building a green roof. I wish I'd had it before building my two green roofs (the first one in 1960). It begins with cautions about leaks and total weight, then lays out the many aspects of what is actually a complex process. It's well-written, well-researched, and full of clear and informative drawings and photos.

—Lloyd Kahn, editor-in-chief, Shelter Publications, author,
Shelter and *The Half-Acre Homestead*

I can't begin to count the number of clients I've had who are interested in incorporating a green roof into their project. Now, thanks to Leslie we have an excellent resource for both DIY and design professionals.

—Tim Krahn, P. Eng. Building Alternatives Inc., author,
Essential Rammed Earth Construction

Leslie Doyle has brought the art and science of green roof construction to the masses. Abundantly approachable, yet thorough and exceptionally detailed, this is a go-to resource for anyone with an interest in building their own green roof. This book is yet another excellent addition to this series — and to my library!

—Chris Phillips, Greening Homes Ltd.

sustainable building **essentials**

essential
GREEN ROOF CONSTRUCTION
the complete **step-by-step** guide

Leslie Doyle

Copyright © 2022 by Leslie Doyle.
All rights reserved.

Cover design by Diane McIntosh.

Cover images supplied by Restorations Garden, Inc. and ©iStock

Printed in Canada. First printing November, 2021.

This book is intended to be educational and informative. It is not intended to serve as a guide. The author and publisher disclaim all responsibility for any liability, loss or risk that may be associated with the application of any of the contents of this book.

Inquiries regarding requests to reprint all or part of *Essential Green Roof Construction* should be addressed to New Society Publishers at the address below. To order directly from the publishers, please call toll-free (North America) 1-800-567-6772, or order online at www.newsociety.com

Any other inquiries can be directed by mail to:
New Society Publishers
P.O. Box 189, Gabriola Island, BC V0R 1X0, Canada
(250) 247-9737

LIBRARY AND ARCHIVES CANADA CATALOGUING IN PUBLICATION

Title: Essential green roof construction : the complete step-by-step guide / Leslie Doyle.

Names: Doyle, Leslie, author.

Series: Sustainable building essentials.

Description: Series statement: Sustainable building essentials | Includes bibliographical references and index.

Identifiers: Canadiana (print) 20210270233 | Canadiana (ebook) 20210270284 | ISBN 9780865719156 (softcover) | ISBN 9781550927108 (PDF) | ISBN 9781771423069 (EPUB)

Subjects: LCSH: Green roofs (Gardening)—Design and construction.

Classification: LCC SB419.5 .D69 2021 | DDC 635.9/671—dc23

New Society Publishers' mission is to publish books that contribute in fundamental ways to building an ecologically sustainable and just society, and to do so with the least possible impact on the environment, in a manner that models this vision.

New Society Sustainable Building Essentials Series

Series editors
Chris Magwood and Jen Feigin

Title list

Essential Hempcrete Construction, Chris Magwood
Essential Prefab Straw Bale Construction, Chris Magwood
Essential Building Science, Jacob Deva Racusin
Essential Light Straw Clay Construction, Lydia Doleman
Essential Sustainable Home Design, Chris Magwood
Essential Cordwood Building, Rob Roy
Essential Earthbag Construction, Kelly Hart
Essential Natural Plasters, Michael Henry & Tina Therrien
Essential Composting Toilets, Gord Baird & Ann Baird
Essential Green Roof Construction, Leslie Doyle

See www.newsociety.com/SBES for a complete list of new and forthcoming series titles.

THE SUSTAINABLE BUILDING ESSENTIALS SERIES covers the full range of natural and green building techniques with a focus on sustainable materials and methods and code compliance. Firmly rooted in sound building science and drawing on decades of experience, these large-format, highly illustrated manuals deliver comprehensive, practical guidance from leading experts using a well-organized step-by-step approach. Whether your interest is foundations, walls, insulation, mechanical systems, or final finishes, these unique books present the essential information on each topic including:

- Material specifications, testing, and building code references
- Plan drawings for all common applications
- Tool lists and complete installation instructions
- Finishing, maintenance, and renovation techniques
- Budgeting and labor estimates
- Additional resources

Written by the world's leading sustainable builders, designers, and engineers, these succinct, user-friendly handbooks are indispensable tools for any project where accurate and reliable information is key to success. GET THE ESSENTIALS!

Contents

Acknowledgments		ix
Chapter 1:	Introduction to Green Roofs	1
Chapter 2:	Green Roof Layers and Roofing Terminology	7
Chapter 3:	Before You Start	13
Chapter 4:	Roof Access and Safety	27
Chapter 5:	Site and Design Factors	35
Chapter 6:	Plants	43
Chapter 7:	Green Roof Material Options	63
Chapter 8:	A Rural New Build	97
Chapter 9:	An Urban Retrofit Build	107
Chapter 10:	Maintenance	115
Chapter 11:	Food Production Roofs	123
Appendix A:	A List of Common Standards and Guidelines	135
Appendix B:	North American Cities with Green Roof Programs	137
Appendix C:	Relevant Climate Links	139
Endnotes		141
Bibliography		145
Index		151
About the Author		157
About New Society Publishers		158

Acknowledgments

To my mom and dad: your words and lessons will never be forgotten.

A big thank you to Chris Magwood for teaching me how to look at buildings through a sustainable lens. Thank you for encouraging me to write this book and for your thoughtful edits and contributions. Thank you to the New Society team for the opportunity and making the book better than I could have ever imagined. Thank you to all the past Restoration Gardens staff (Keara White, Steve Massey, Sarah Rafols, Anna King, David King, Grayson Sherritt, Yolanda Lloyd, Jesse John, and Katie Howard) for your ideas, heavy lifting, and research on all our wonderful projects and for capturing them so beautifully in the pictures used throughout these pages. Thank you to Femke Bergsma of Grame; Arlene Throness, Jayne Miles, and Jessica Russell at Ryerson University; and Marc Boucher-Colbert of Urban AG Solutions for taking the time during these challenging days to educate me about edible roof gardens. Thanks go to all of the following: to my friends and family who generously offered quiet spaces, family support, and active listening; to Lorna for raising my children throughout this journey; to my children, Sullivan and Frances, for the renewed inspiration; and to my husband Jackson, for his endless motivation and patience.

Chapter 1
Introduction to Green Roofs

Our roofs, which keep us safe and dry, are subject to a range of environmental stresses. They are exposed to substantial rains and damaging hail, destructive heat and UV radiation, strong winds, and heavy snow. Our changing seasons and associated temperatures can wreak havoc on the waterproof membrane material used on many roofs, causing it to expand, contract, dry, crack, and then leak or break down.

Investing in a quality roofing membrane gives homeowners peace of mind in keeping their house protected. But why not go further? Why not utilize the roof space as a means to contribute to the biodiversity in your neighborhood and offer a protective place for pollinators or birds to feed? Why not reduce the impact of this space by reducing the heat it would normally reflect or capturing the rain that would run off of it? Why not do *all* these things? Why not make your roof do *more*?

You can. You can build a green roof.

Green roofs, eco-roofs, vegetative roofs, vegetated roof assemblies, or living roofs—regardless of the name, they are all roof systems designed to support plant life. Green roofs provide a range of benefits for your site and your community, and they are an opportunity to turn a static surface into a buzzing tapestry of color.

Green Roof Systems

While some green roofs can be complex, others can be quite simple in design. Regardless of your design, for a green roof

Fig. 1.1: *A roof structure gets battered from many exterior forces while protecting precious interiors.* Photo Credit: Restoration Gardens, Inc.

Fig. 1.2: *Green roofs serve many purposes and bring life to an otherwise inanimate roofscape.* Photo Credit: Restoration Gardens, Inc.

Fig. 1.3: *A built-in-place system is installed in layers on the roof.* Photo Credit: Restoration Gardens, Inc.

Fig. 1.4: *An example of a hybrid system: layers are installed individually and finished with pre-grown vegetation mats.* Photo Credit: Restoration Gardens, Inc.

to succeed it must include the following five main elements:

- a strong roof structure
- a waterproof membrane
- suitable growing media
- good drainage
- appropriate plants

Understanding how these elements come together will allow you to design and build the green roof you need. Throughout this book, I will provide you with visual examples to guide you.

You can build a green roof layer by layer (a built-in-place [BIP] system), or you can install manufactured products, such as trays that come pre-fitted with all the needed layers, often including pre-grown vegetation. You can also create hybrid systems. For example, layers can be installed on the roof with a pre-grown vegetation mat unrolled overtop, or you can install modules pre-filled with media and complete the planting on the roof.

Green Roof Categories

Green roofs fall into three categories: *Extensive, Intensive,* or *Semi-Intensive.*

Extensive systems have shallow planting, making them inhospitable to many species. This means they support less plant diversity; however, they require less maintenance and upfront cost. These are the types of roofs most people think of when they think "green roof." *Intensive systems* have few restrictions and can be designed as accessible public parks with soil depths deep enough to support trees. The middle ground of a *semi-intensive* roof typically allows for greater plant diversity than an extensive roof and therefore requires a little more maintenance, but they are not as resource heavy as intensive roofs. See Photos 1–3 in the Color Section for images of each kind.

This book is focused on extensive roofs and semi-intensive roofs, but it also introduces food production roofs. Food roofs, rooftop farms, and edible roofs are all terms referring to green roofs that are built or modified for the production of food. These are often considered semi-intensive or intensive roofs due to the depth of the growing media and the frequency of maintenance that is required. Photo 4 in the Color Section shows a food roof in an urban setting.

Extensive, semi-intensive, and small food roofs can easily be built and managed by an enthusiastic and competent DIYer on simple residential or outbuilding rooftops. Intensive roofs should be left to professional contractors, as they require an integrated approach with input from architects, engineers, landscape architects, and material suppliers. The table in Figure 1.5 highlights comparisons between the three types of green roofs this book will cover.

New Builds versus Retrofits

Green roofs can be built on both new and existing roofs. On new roofs, you have the freedom to build the roof structure based on your green roof design intent, whereas on an existing roof, your green roof may not exceed the existing structural capacity. If your

Figure 1.5: Comparisons of Green Roof Categories and System Types

	Extensive	Semi-Intensive	Food Production Roof
Depth of Growing Media	<6"; the ideal minimum is 4"*	Approximately 4–8"	Varying depths of 6–12" or higher
Type of Growing Media	Traditionally aggregate-based with low percentage of organics	Typically involves a higher organic content	Can be as much as 100% organic
Typical Plants	Moss Sedums Some grasses Some flowering perennials	Grasses Flowering perennials Some bulbs Some shrubs	Fruits, vegetables, and herbs
Irrigation Needs After Establishment	Low	Moderate; depends on soil composition and climate	High
Modular Systems Available	Yes	Yes	Can use some modular components but do planting yourself
Saturated Weight**	4" systems are approximately 20–30 psf	30–60 psf	50–100 psf, depending on the depth of soil over 6"
Cost***	Low	Moderate	Most expensive (in terms of ongoing resource and maintenance needs)
Maintenance	Lowest	Moderate	High

* Systems with less than 4 inches of growing media tend to require permanent irrigation systems and provide fewer ecosystem services. Systems with shallow growing media are most appropriate for existing roofs with restricted weight loads.
**Ranges are based on commercially available blends. Custom soil blends could be higher.
***Total costs typically escalate as the growing media deepens. On a sq. ft basis, the cost per sq. ft will go down the larger the roof gets. For a DIY install (where labor is typically free), you will have to do your own cost comparisons for materials to compare modular and BIP systems.

existing roof does not allow for the added weight of a green roof, you can retrofit your roof to accommodate your loading needs; however, this can be a costly venture.

This book will give you everything you need to know about building a green roof but it does not directly show you how to design or build the roof structure underneath it. The typical green roof assembly does not include the roof structure itself. However, these two building components are highly interdependent. Reviews of local codes and/or a call to a structural engineer are necessary to ensure your structure is adequate for the amount of additional loading you intend to add.

How This Book Can Help You

The green roof industry is still relatively young and, therefore, still evolving. In North America, awareness and interest in green roofs has exponentially increased in the past decade and a half, and there is a growing body of research and publications. Whereas the early research was concerned with how the newly developed systems from Germany could be of use to our cities, research is now more focused on region-specific systems and designs for specific ecosystem services. Twenty years ago, the only books published on green roofs were intended to educate contractors on this new type of construction. Now, many books are published filled with award-winning designs of green roofs around the world. In addition, the industry is still learning many valuable lessons and developing products specifically for green roofs. As awareness and interest increase, many individuals are left wondering how they might build their own simple system.

Referencing lessons from these valuable resources as well as my experience in building green roofs for the past ten years, this book aims to provide the necessary information and planning steps for the Do-It-Yourself builder. The lessons in this book are for homeowners building simple green roof projects; they are not intended to be applied to complex installations or used by commercial installers (who must strictly adhere to municipal building and safety codes and may have to follow manufacturer recommendations for warranty purposes).

My goal for this book was to use common language that would not require a horticulture or construction background; however, I want you to be familiar with the key terms so you feel competent when ordering supplies or talking to professionals.

In addition to terminology, you will find planning steps, installation and design considerations, plant and material choices, two installation examples, maintenance considerations, and an introduction to food production roofs. Consideration is given to the diversity of North American climates. Many details are provided so that regardless of the size of your build, you will have all the necessary planning preparation in front of you. However, if you are merely building a small green roof over a tool shed or a dog shed (in no way am I diminishing Spot's need for well-being), you may not need to consider all the steps found in Chapters 3 and 4. They are still worth a read, however, so you can make informed decisions down the road.

Attention to Details

Racing through details can cause big problems, especially when you are dealing with water overtop of your home. Do not underestimate the ability of water to find small imperfections. In this book, you will find very detailed drawings illustrating the concepts; please review these carefully and apply the same level of detailed attention to any new ideas or variations you may wish to perform.

Fig. 1.6: *Water is starting to find a way through this green roof after 10 years.*
Photo Credit: Restoration Gardens, Inc.

Chapter 2

Green Roof Layers and Roofing Terminology

Green Roof Layers

A GREEN ROOF INCORPORATES a series of layers that work together to protect the building underneath it from water while supporting plant life. These layers are divided into two categories: (1) those that are required to ensure the green roof will function properly, and (2) those that are optional based on the design specifics. Descriptions of these layers are found in this chapter, and in Chapter 7 you will find material specifications.

Required Layers

Waterproof Membrane

The most essential part of any green roof, a waterproof membrane keeps water out of the structure, protecting the integrity of the building. This layer usually lies directly on top of the building's roof deck.

Root Barrier

A root barrier protects the membrane from root penetration. Some roots are quite aggressive and can cause damage if allowed

Fig. 2.1: *Required layers of a green roof.*

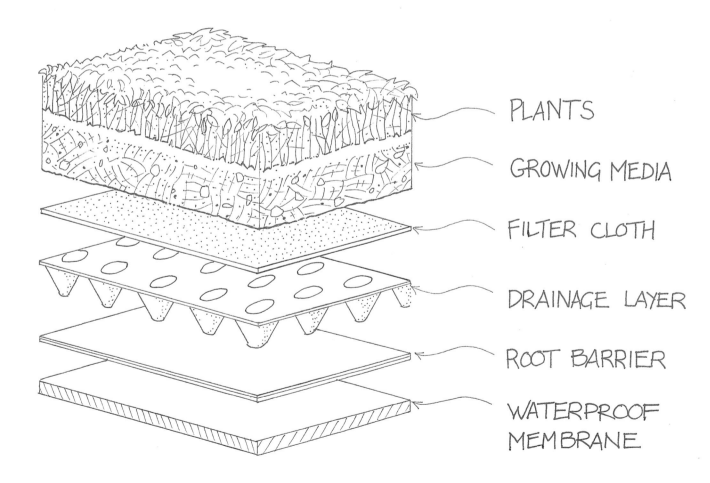

to grow without restrictions. This layer is an absolute requirement if the membrane does not contain root-resistant properties.

Drainage (Layers and Outlets)

The drainage layer allows water to flow freely through the built assembly and toward the drain outlets. Drainage outlets allow water to run off the roof. Both drainage elements (down and out) are required to eliminate standing water (especially on flat roofs).

Filter Cloth

Also known as geotextile, filter cloth is required to keep the fine particles of the growing media out of the drainage layer, ensuring that the drainage layer does not become clogged, which would restrict the movement of water.

Fig. 2.2: *Roots growing within the granules of a typical modified bitumen roof.*
Photo Credit: Restoration Gardens, Inc.

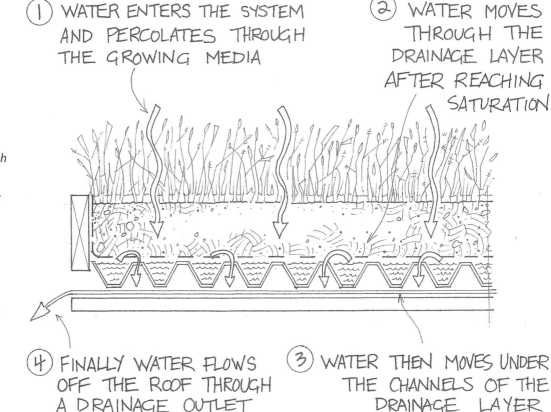

Fig. 2.3: *Drainage systems allow the passage of water through the growing media to mitigate oversaturation; they also give water an exit from the roof to reduce ponding and excessive weight. Some drainage layers are designed to retain small amounts of water for plants to use, but allow most of the water to move through.*

Growing Media

Growing media is the substrate in which the plants grow. As opposed to most gardening soils, media on a green roof is typically a mix of organic material and lightweight aggregate.

Plants

Plants are the last thing to go on the green roof. They serve as visual indications of how well you have built the assembly and the health of your system.

Optional Layers

In addition to the required layers of a green roof, there are optional layers. The necessity of these items will be based on your green roof design.

Membrane Protection

This layer protects the membrane from continuous friction or pressure points coming from the overlying layers. Membrane protection is normally required for intensive systems because they are heavier and are more prone to foot traffic; the need for membrane protection in extensive systems depends on your membrane type and the overlying drainage layer.

Insulation

Insulation in a green roof is typically in the form of rigid boards. In flat roofs, insulation can be part of the underlying roof structure as the primary thermal barrier, or it can be added on top of the roof structure, where it can act either as the primary thermal barrier or as an enhancement to the thermal performance of insulation that already exists in the ceiling.

Water Retention

A water-retention layer temporarily holds water in between rain events. A water retention layer may be required for green roofs with steep slopes, in areas with high winds, or for roofs with very shallow depths (i.e., 2 inches or less).

Slope Restraints

On steep green roofs, a manufactured or built-on-site restraint system must be incorporated to prevent growing media from sliding.

Wind Erosion Protection

This type of protection is a temporary or permanent mechanism or design feature intended to prevent wind from blowing the growing media off the roof.

Irrigation

It may be necessary to provide a temporary or permanent source of supplementary water to plants in between rain events.

Roof Terminology

Gable Roof

A roof with two sides that meet at a ridge.

Shed Roof

A simple roof with one slope. This is probably the best type for a DIYer to work with the first time.

Roof Deck

The roof deck is the construction material that sits between the structural supports (joists or trusses) and the waterproofing material (membrane). In residential applications, it usually consists of plywood or tongue-and-groove lumber.

Parapet

A parapet is a small wall that runs along the perimeter of the roof deck. In a green roof application, the parapet works to keep the green roof material contained on the roof as well as aid in wind erosion protection. It can be made of stacked or formed lumber sitting on top of the roof deck, or it can be made of stacked stone; it can also be a continuation of the exterior wall.

Cant Strip

A triangular strip placed in the angle between a roof and a parapet or any other vertical surface to which the roof abuts (including building walls). This strip provides support behind membranes in areas where sharp corners may be hard to achieve. Cant strips can be made of concrete, wood, steel, insulation, or insulation composites.

Scupper

A scupper is an opening in the parapet where water exits. It is usually at the lowest point of the slope, and more than one may be necessary based on the slope, design, or size of the roof.

Emergency Overflow

An emergency overflow is a fail-safe way to ensure that water can escape from the roof in the event of a clogged scupper. In a situation where the parapets are high, a clogged scupper would result in excess water pooling—potentially exceeding the weight allowance for the structure.

Cricket

A cricket is a high point designed into the roof deck to divert water. They are often used around building structures, such as chimneys, and at transitions from one roof section to another; they are also used on roofs where one slope may not clear all the water off the roof.

Fig. 2.4: *The placement of a cant strip can make angles easier to work with.*

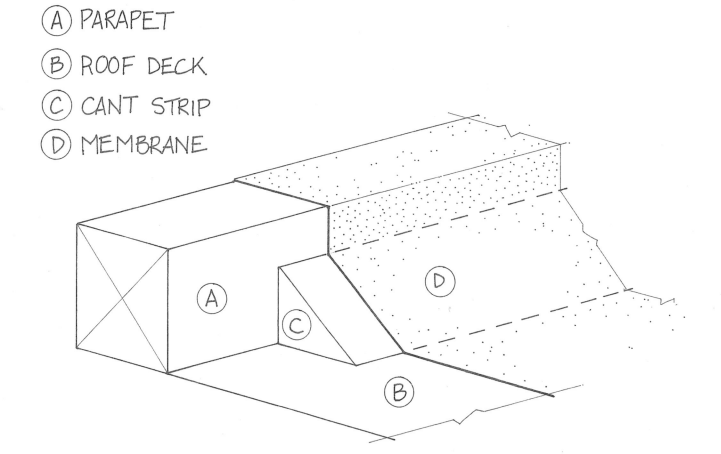

Drip Edge

A drip edge is a metal flashing used to force the water off the lower edge of the roof instead of allowing it to trail back toward the house. It is also the term used for the lower edge of the roof where water drains off.

Rake Edge

The rake edge runs along the roof from the lowest point to the highest point. Water typically does not flow off these edges.

Fig. 2.5: *A flat roof with a typical scupper. Scupper openings allow for water to exit the roof and into a downspout.*
PHOTO CREDIT: RESTORATION GARDENS INC.

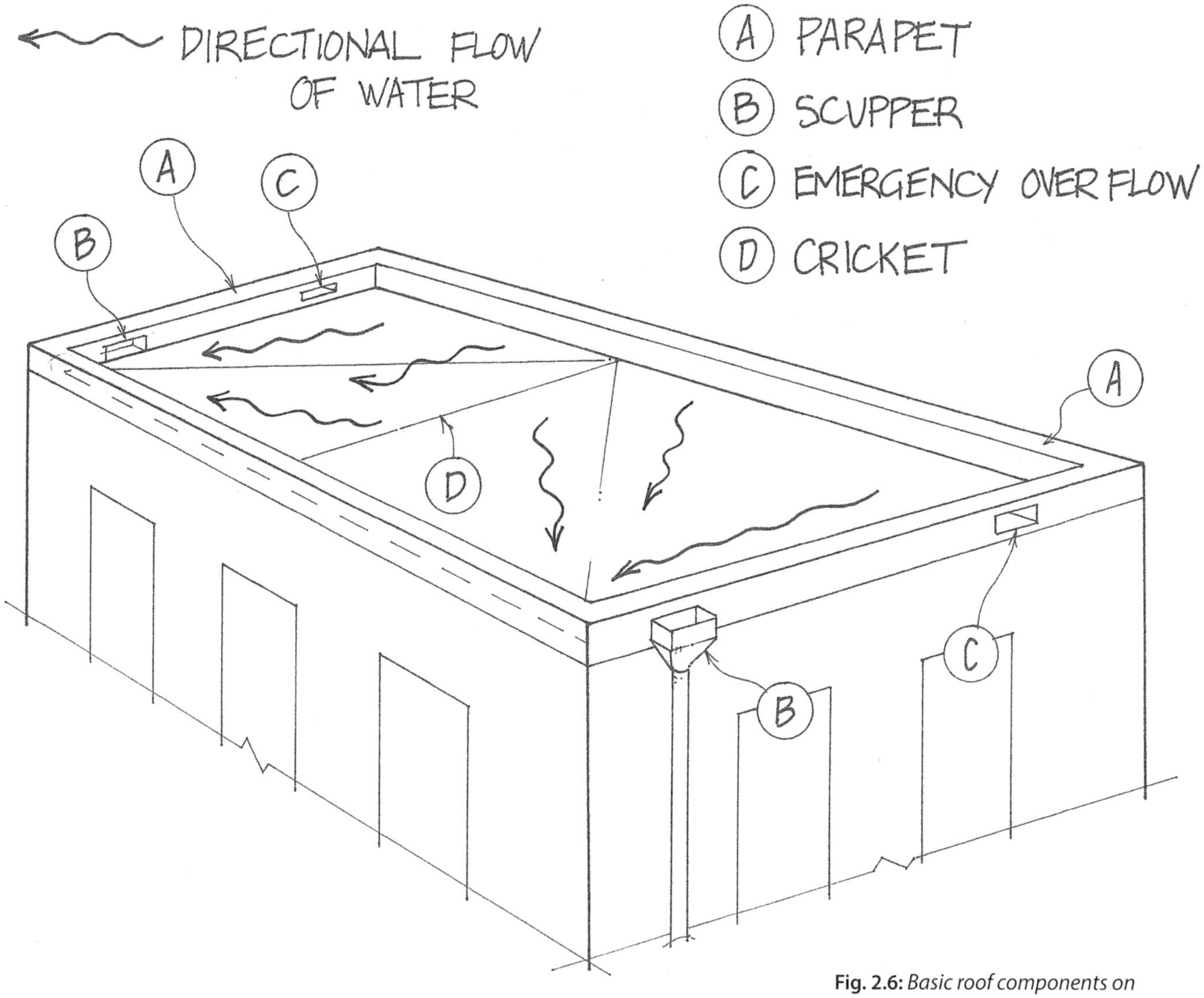

Fig. 2.6: *Basic roof components on a typical flat roof with parapets.*

Eaves

Eaves are the section of the roof that overhangs from the wall out to the end of the joist. The job of the eaves is to protect the walls from water coming off the roof and/or from rain.

Gutter/Eavestrough

Gutters, as they are known in the US, or *eavestrough*, as they are known in Canada, are channels that are fastened to the fascia board at the drip edge. Gutters collect water and direct it down a downpipe (or rain chain) away from the foundation of the house.

Now that you have a basic understanding of a green roof system and the parts of a roof, we can go through how to build one.

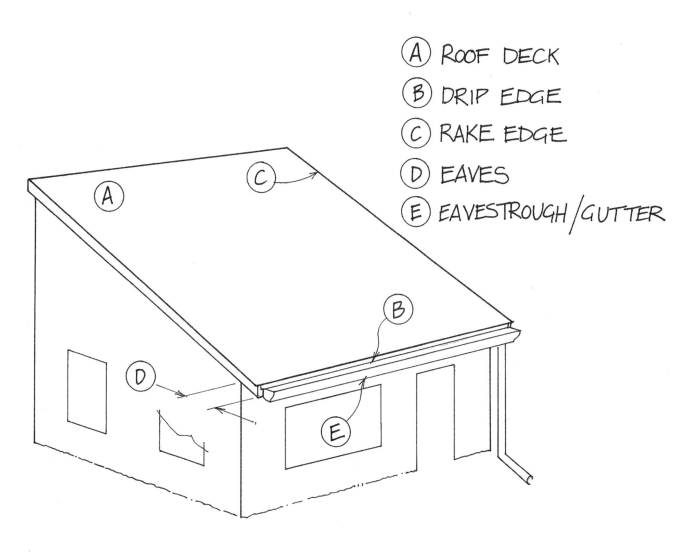

Fig. 2.7: *Basic roof components on a typical shed roof with no parapets.*

Chapter 3

Before You Start

There are many reasons why homeowners choose to build a green roof. Besides the unrivaled beauty they offer over traditional flat roofs, green roofs serve many purposes on a small scale, such as increased thermal and sound insulation, reduced site run-off, increased amenity spaces, and an increased membrane lifespan. Although most green roofs can serve multiple purposes quite easily, it is important for you to be clear about what your specific goals are for your roof.

Fig. 3.1: *The green roof design goals for the bermed house were to blend into the natural surroundings, provide insulation, and replace green space lost by the build.* Photo credit: Neil Hodder

The table in Figure 3.2 outlines some design goals with corresponding considerations.

While many people get excited over the type of green roof they want to build, the first step will be to establish if it's possible. Start with these steps:

- Check with your local municipality for any regulatory requirements.
- Check if your insurance policy coverage will need to change.
- Have a structural engineer determine if your roof can handle the extra loading.

If you are building a new structure, you have greater flexibility, as you can build the roof to handle the extra loading required.

Codes and Standards

Standards for the construction of a green roof assembly are not yet typically part of building codes. However, the membrane and any roofing insulation you include will have to adhere to local code regulations as it relates to the roof structure.

Before starting your green roof planning, review your municipal bylaws or regulations. This will determine if your municipality requires the construction of the green roof to follow a set of standards or guidelines (as mentioned above) and will determine if you qualify for any local incentives. Appendix A gives a list of some common standards and guidelines that may have been adopted by your local municipality. Appendix B contains

Figure 3.2: Common Design Goals and Considerations for Homeowners

Design Goals	Considerations
Aesthetic-Based	
Improve the View	Minimal loading, extensive system. Focus on lightweight products and low maintenance plantings.
Increase Biodiversity	Higher loading, semi-intensive system. Increase planting depth and organic content of growing media. Some species require more water, so consider water-retention layers to avoid having to irrigate.
Function-Based	
Reduce Stormwater Run-Off	Higher loading, semi-intensive system. Growing media components should be chosen based on water-retention capabilities. Consider moisture retention mats and plants which have higher rates of transpiration.
Increase Insulation Value of the Roof	Extensive or semi-intensive systems. Add suitable insulation for system components. Include some growing media components that keep the roof cooler (i.e. vermiculite) and plants that offer more shading of the surface.
Produce Food	Higher loading, semi-intensive system. Increase water retention and/or irrigation. Increase organic content. Be aware of increased safety and access needs.
Other Design Goals	
Low Maintenance Roof/Lightweight Green Roof	Minimal loading, extensive system. Use shallow inorganic substrates to support mainly succulent species.
Green Roof in the Subtropics	Extensive or semi-intensive systems. Growing media should balance drainage with water retention to handle seasonal flooding and periods of drought. Incorporate vermiculite to retain water but reduce growing media heat transfer during the summer. Choose plants adapted to regional stressors that can provide some canopy shading.[1]
Green Roof with a Low Carbon Footprint	Extensive or semi-intensive system. Minimal loading to reduce underlying joist size. Focus on local and/or recycled materials or choose materials that can be recycled at end of life. Choose soils with higher levels of organics (which could include biochar and/or compost). Plan for diverse, competitive, drought- and stress-tolerant plantings that produce more biomass than succulents, for increased CO_2 absorption.

a list of North American cities that have green roof programs in place (as of early 2021).

For example, in Toronto, there is a bylaw that requires and governs the construction of green roofs on buildings over 21,500 square feet (2,000 m²) in gross floor area.[2] While this does not include home projects, the City of Toronto does require permits for all green roofs in the City and requires that they conform to the Toronto Green Roof Construction Standard. The City provides a supplementary guideline with illustrations offering best practices for achieving the Standard. For those who voluntarily build a green roof, the City offers a financial incentive for both the initial structural assessment and the installation. The City also offers great guidelines for creating green roofs that build habitat and promote biodiversity.

Be sure to read standards and guidelines carefully. Some may require a designated Green Roof Professional (GRP) or other registered design professional to submit the applications, provide drawings, or build the roof. Take special note that grants usually require you to start the process of application *prior* to starting construction.

Even if your jurisdiction doesn't have regulations or a set of construction standards, you will still want to know about the topics covered in this chapter. For a simple green roof built by a homeowner, it is important to understand how a green roof system works, the role of each layer, how to choose appropriate materials, and what is necessary to maintain the integrity of the structure and waterproof membrane.

Insurance

Some home insurance policies may change based on the addition of a green roof. Be prepared to present your insurer with your intended risk-mitigation strategies such as adequate structural design, roof details, fire prevention, and maintenance. With the escalating costs associated with flood damage in many cities, insurance companies are starting to welcome low-impact design solutions.

Structural Loading Capacity

Structural loading capacity is one factor that determines whether or not a green roof is possible. There are two loads to consider when planning a green roof:

- **Dead Load:** The weight allowance for all permanently placed material. This includes the roof structure and the green roof assembly (including plants and the weight of the amount of water required to saturate the growing media), in addition to any hardscaping or permanent fixtures.
- **Live Load:** The weight allowance for all temporary material. This includes weights imposed by weather such as wind, rain, and snow, as well as temporary items such as equipment or people. Live load allowances vary by region.

You should have full confidence that your roof is built to meet (or exceed) all weight load requirements. If you are doing new construction, design your walls and roofs with sufficient strength and stiffness to support your desired green roof. I always recommend *over*-building your structure to leave you with some flexibility, should you need it. Always refer to span tables in your local building codes.

Weight loads are provided in pressure conversions. In this book, I will be using *pounds per square foot* (psf). Refer to Figure 3.3 for a list of common weight allowances found on roofs and the corresponding class of green roof allowed.

For existing roofs, you can assume that your roof meets the minimum regional requirements, but you will require a structural engineer (or other registered design professional) to let you know what dead load capacity, if any, is available for the addition of a green roof. If you are provided with a low weight value (~15 psf) from an engineer, you can still create a green roof, but you will be limited in the depth of media and, therefore, the plants you are able to grow.

Those who currently have a flat roof with gravel ballast are likely able to replace the ballast with a very light green roof system, as ballast roofs typically weigh 10–15 psf (48.8–73.2 kg/m²).[3] However, this should be confirmed first. A structural engineer can look at existing permit drawings, or they can inspect the roof to determine the joist size and spans for their calculations.

When you calculate weights of materials, you must consider *saturated* loads, as most materials are designed to absorb water. A product that retains 0.15 gal/ft² (6 L/m²) would add approximately 1.2 psf (5.86 kg/m²) to the load, as the weight of 1L of water is the equivalent to 1kg of water.

If you need to provide a weight for permitting, look to materials that are available to you and gather the necessary weights from the manufacturers. You can also find some weights in Chapter 7. Growing media weights can be harder to calculate. FM Global recommends using a saturated weight load of 100 lb/ft³ (1601.85 kg/m³) for growing media.[4] This would result in the growing media layer weighing 33.33 psf (162.73kg/m²) at 4 inches, and 50 psf (244 kg/m²) at 6 inches. This may be a safe calculation if you are purchasing an engineered mix, but if you are blending your own, it may be too low. See Chapter 7 for instructions on performing your own saturated weight load tests.

There are many products and systems available with varying loads. Figures 3.4 and 3.5 offer examples of some weight loads based on some available materials for extensive and semi-intensive roofs.

Figure 3.3: Common Weight Allowances and Units

Green Roof Class	Pressure Conversions			
	Imperial		Metric	
	Pounds per square inch (psi)	Pounds per square foot (psf)	Kilopascals (kPa)	Kilograms per square meter (kg/m²)
Extensive	0.1451	20.89	1.00	101.97
Extensive	0.2083	30.00	1.44	146.47
Extensive/Semi-Intensive	0.2901	41.77	2.00	203.94
Semi-Intensive/Food Roof	0.3472	50.00	2.39	244.12
Semi-Intensive/Food Roof	0.4351	62.66	3.00	305.93
Food Roof/Intensive	0.6944	100.00	4.79	488.24

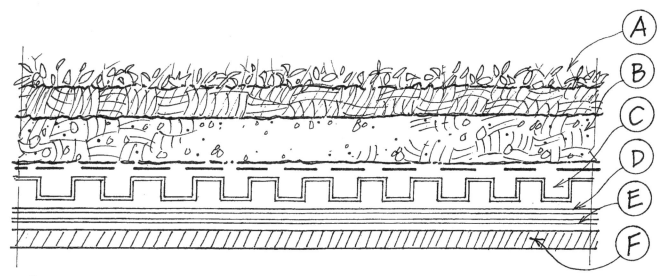

Ⓐ PRE-VEGETATED SEDUM MAT, 1" (2.5cm) 2.25 psf [5]

Ⓑ GROWING MEDIA (SAND, GRAVEL + TOP SOIL MIX) 1" ≃ 10 psf

Ⓒ DRAINAGE BOARD w/ RETENTION FLEECE, 3/8", 1.36 psf [6]

Ⓓ EPDM MEMBRANE (ROOT RESISTANT), 0.43 psf

Ⓔ GEOTEXTILE AS PROTECTION SHEET, 0.04 psf [7]

Ⓕ ROOF DECK, NOT INCLUDED IN CALCULATIONS

SATURATED WEIGHT LOAD = 14.08 psf

Fig. 3.4: *To achieve a very lightweight system, plan for about 2 inches of substrate depth to support* Sedum *species.*

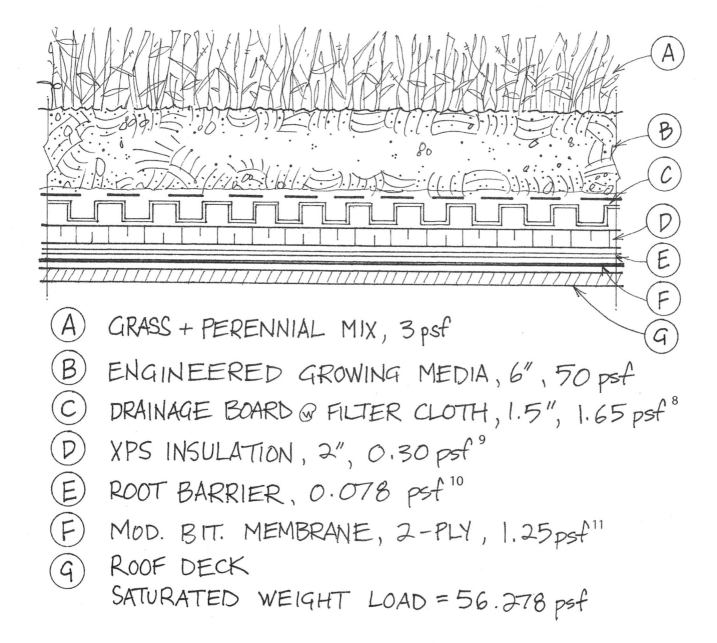

Fig. 3.5: A semi-intensive green roof system with 6 inches of growing media and 2 inches of supplementary insulation.

Point Loading

A structural engineer can provide you with a total design load, but you should also ask for any point loads. These are specific areas on the roof with underlying columns or structures that can handle more weight. Knowing where these areas are gives you the opportunity to create more interest on the roof, such as deeper growing media features for increased plant diversity.

Once you know the structural loads of your roof, you can explore your options for:

- Occupancy/access
- Function
- Growing media composition and depth

- Plant selection
- Installation strategies
- Replacement and repair strategies

Occupancy/Access

Green roofs on residential buildings can provide great places of enjoyment. Many urban green roof owners build a green roof as part of a rooftop patio that includes shade structures and raised planters. If this is the intent for your design, please review your local building codes, as these roofs are considered occupied spaces and will require a greater live load, specific safety measures (such as railings), and possibly rezoning. Access for maintenance purposes only does not require additional loads and rezoning, but safety measures should be in place.

Fig. 3.6: *Finding point loads on your roof can allow you to create areas for heavy materials and activities such as composting.* Photo Credit: Leslie Doyle

Fig. 3.7: *A roof designed for occupancy with railings made of glass and cedar.*
Photo Credit: Restoration Gardens, Inc.

Fig. 3.8: *A large window can be used for access.* Photo Credit: Restoration Gardens, Inc.

Safety concerns for occupants center on access to and exit from the roof, as well as keeping everyone safe while on the roof. Safe-access options include exterior staircases, permanent ladders, patio doors, large windows, or roof hatches.

Your local code will dictate railing heights and other variables; in Ontario, for example, railings must be 42 inches (106.68 cm) high with openings of less than 4 inches (10 cm), non-climbable and they must wrap around the perimeter of the accessible area. The height is to be based on the finished height of the green roof growing media. (Note that parapets can act as railings if they meet the required height. Otherwise, you can attach railings to them to make up the difference.)

Fig. 3.9: *These railings, fixed to the parapet, create a safe railing height of 42 inches.*
Photo Credit: Restoration Gardens, Inc.

Function

Green roofs can be designed to serve different functions including, but not limited to:

- Stormwater retention
- Increased insulation value
- Food production

Such functions will affect and likely increase the loading needs for the green roof. Stormwater retention designs may include a deeper drainage layer, deeper growing media, or growing media that retains or absorbs moisture for long periods of time. Using additional growing media versus rigid insulation for added insulation value can increase the weight load. Food-producing roofs require deeper growing media and require occupancy on the roof, both of which increase the loading needs of the roof.

Growing Media Composition and Depth

The load capacity of your roof will affect the composition and depth of growing media you will use. Commercial suppliers offer a range of engineered growing media to suit different load allowances. You can also mix your own, but it is a bit tricky to get right. I cover this later, in Chapter 7.

Plant Choices

Plant diversity is related to the growing media composition and depth. Roofs with lighter load allowances tend to support only succulents, moss, and a few species of flowering perennials. As the substrate deepens, the diversity of plants can increase. See Photo 5 in the Color Section for a roof that supports many different species due to the different

Fig. 3.10: *Roofs designed for food production and frequent visitors need additional weight accommodations.*
PHOTO CREDIT: RESTORATION GARDENS, INC.

depths of growing media incorporated into its design.

Installation Strategies

Installation strategies are the plans you make for getting materials onto the roof and staged. Normally, on a residential roof, I advise against staging too much on the roof because it limits your ability to lay down materials efficiently; it is better to hoist materials up as needed. Bringing materials up only as needed also ensures you are not exceeding weight loads. If staging on the roof is necessary, place material along the perimeter or near the walls; that is where you have the highest loading, due to the underlying structure. On larger projects, where a crane is required to get materials onto the roof, get everything up and staged in an organized fashion in one day, as it would be too costly to hire the crane operator every time you need to lift material. An organized plan based on known point loads can make for an efficient install. Sometimes, in rural areas, a neighbor with a front end loader may be your best friend.

Replacement and Repair Strategies

No one wants to have to repair or replace a green roof. One of the benefits of a green roof is that, if built correctly, it will outlast a traditional flat roof. However, there may be a time when a replacement or repair needs to occur. These situations may include:

- Renovations to the interior or exterior of the home that require new penetrations through the roof, such as vents or skylights or new walls.
- Improper waterproofing that requires the green roof layers to be temporarily pulled back.
- Improper flashing.
- Plant death due to improper maintenance.
- A missing essential layer.

The point of this book is to help you avoid problems so replacements or repairs do not need to occur. However, no one can predict if future interior renovations will be desired. If repairs are needed, knowing the weight allowances on the roof allows you to strategically place the overburden (the materials and layers) that need to be temporarily moved. Consideration of the weather is an important factor. Work should be done when rain or snow is not in the near forecast, as this can saturate your staged material.

Fig. 3.11: *If possible, bring materials up only when they are needed so you can install layers without having to work around staged materials.*

PHOTO CREDIT: RESTORATION GARDENS, INC.

Photo 1, top left: *An extensive roof.*
PHOTO CREDIT: RESTORATION GARDENS, INC.

Photo 2, top right: *A semi-intensive roof.*
PHOTO CREDIT: RESTORATION GARDENS, INC.

Photo 3, below: *An intensive roof.*
PHOTO CREDIT: RESTORATION GARDENS, INC.

Photo 4, above: *A food roof.* PHOTO CREDIT: RESTORATION GARDENS, INC.

Photo 5, right: *4-inch depths support succulent species while the diversity increases at 6–8 inches. This roof supports trees and wildflowers around the 8" perimeter as well as in the large planters.* PHOTO CREDIT: RESTORATION GARDENS, INC.

Photo 6 and 7, top and center left: *This roof required a diverse planting plan to accommodate the various sun exposures;* Sedums *and* Allium schoenoprasum *grow in the shallow sun, while ferns, shade-tolerant sedums,* Allium cernuum *and* Aquilegia canadensis *grow in the shade with deeper substrate.* PHOTO CREDIT: RESTORATION GARDENS, INC.

Photo 8: *An alvar community in Eastern Ontario blooming with* Solidago ptarmicoides *(upland white goldenrod).* PHOTO CREDIT: LESLIE DOYLE

Photo 9: *This planting plan designed by Natvik Ecological, consists of predominantly 4 inches of growing media with 6-inch mounds over loading points allowing for a mix of* Sedum *and native perennials. Clockwise from top left:* Sedum reflexum *'Blue Spruce';* Sedum kamstchaticum; Allium schoenoprasum; Bouteloua curtipendula; Sedum spurium; Sedum acre, Achillea millefolium *'Paprika';* Nepeta *'Walker's Low'.* PHOTO CREDIT: RESTORATION GARDENS, INC.

Photo 10, above: *Allium schoenoprasum (common chive) is a prolific self-seeding perennial, and it's edible too!* PHOTO CREDIT: RESTORATION GARDENS, INC.

Photo 11, center: *A green roof over an earth shelter home planted with Eco-Lawn grass seed mix.* PHOTO CREDIT: NEIL HODDER

Photo 12, below: *Moss Acres in Pennsylvania sells moss for green roofs and demonstrates here how it can grow on a green roof.* PHOTO CREDIT: LESLIE DOYLE

Photo 13, above left: *Moss mixed with hydrogels and placed on a biochar growing media on green roof test plots in Lindsay, Ontario.*
PHOTO CREDIT: LESLIE DOYLE

Photo 14, above right: *Two years after spreading the cuttings, they have formed a lush mat with full coverage.*
PHOTO CREDIT: RESTORATION GARDENS, INC.

Photo. 15, left: *This roof showcases four ways to plant Sedum. Bottom to top include modular trays, pre-vegetated mats, plugs, and cuttings.*
PHOTO CREDIT: RESTORATION GARDENS, INC.

16a

16b

16c

Photo 16a, b, c: *This roof was designed with biodiversity in mind. There are two different types of growing media installed, both for drainage services and for insect habitat. Trees provide shelter while logs and rocks hold down wind erosion blankets and create microclimates. The roof was planted primarily with native seeds. The growing media under these rocks were cool and moist even when the rest of the site was dry.* Photo Credit: Restoration Gardens, Inc.

Photo 17, above: *A sloped green roof section includes pre-vegetated Sedum mats to reduce surface erosion as well as wind erosion.* PHOTO CREDIT: RESTORATION GARDENS, INC.

Photo 18, center left: *This roof utilizes the open grate edge with composite lumber.* PHOTO CREDIT: RESTORATION GARDENS, INC.

Photo 19, below left: *Plants have replaced the erosion blanket two years after planting.* PHOTO CREDIT: RESTORATION GARDENS, INC.

Photo 20, below right: *Arlene and Jayne show the difference between their 7-year-old amended growing media mix and newly installed aggregate mix. As told by Arlene, the secret to their success lies in their highly fertile soil.* PHOTO CREDIT: LESLIE DOYLE

Photo 21: *A green roof, accessible by a bedroom window, contains chives, lavender, thyme, and strawberries mixed with other flowering perennials.*
PHOTO CREDIT: RESTORATION GARDENS, INC.

Photo 22, center right: *A fully mature modular tray by LiveRoof Inc.* PHOTO CREDIT: KEES GOVERS

Photo 23, bottom left: *A sloped install built with 4" LiveRoof modular trays.* PHOTO CREDIT: KEES GOVERS.

Photo 24, bottom right: *This roof is composed of materials without any plastics and uses site soil. A grass seed was originally planted 10 years ago, but now the roof cycles through annual grasses, hawkweed and moss.*

Repairs Happen

Example 1: A roofing company on a project of ours failed to make tight corners on the inside of the scupper, and it wasn't until the winter that we learned ice was making its way into that vulnerable point. We worked together to ensure only a small portion of the roof was pulled back and replaced.

Example 2: Welding sparks damaged a section of a completed green roof and caused a leak. The roof had to be opened up, repaired, and then replanted. In this case, because they knew where the interior problem was, the section to be opened was easily found by contractors. We displaced the overburden evenly around the opening so as to not overload sections of the roof while the repair took place.

Example 3: Upon completion of the green roof, a client was awarded grant funding for solar panels, which they wanted to install over the green roof. They also installed an elevator to access the roof. In this case, the construction trades removed the overburden and staged it all safely near the edge of the roof. The project was completed quickly so that we could restore the layers before the added weight of snow blanketed the overburden.

Example 4: We were called to perform maintenance on a roof, and, upon investigating the type of system that was installed, we discovered that they did not have a root barrier—a layer that was not well understood at the time of their install. As the entire roof had to be recovered, we could not store all the growing media on the roof without overloading the structure. Because this roof was in a dense urban setting, it took a considerable amount of effort to find ways to remove the growing media while salvaging the plants and drainage boards.

Slope

Once you know your roof can hold the weight, or you are working with an engineer to design a stronger roof, you should then look at the slope of your roof. The slope of your roof dictates which materials and system components are necessary in order to retain moisture, reduce erosion, and ensure successful plantings.

Determining your Slope

Slope is defined as *rise over run*. If your slope is unknown, climb on your roof with a 12 inch (30 cm) or longer level and a tape measure. On the roof, hold the level perfectly level, and measure the height from the roof to the level 12 inches away from where the level touches the surface. This will be the rise (see Figure 3.12). For example, if the level is

Fig. 3.12: *An easy way to find the slope of your existing roof.*

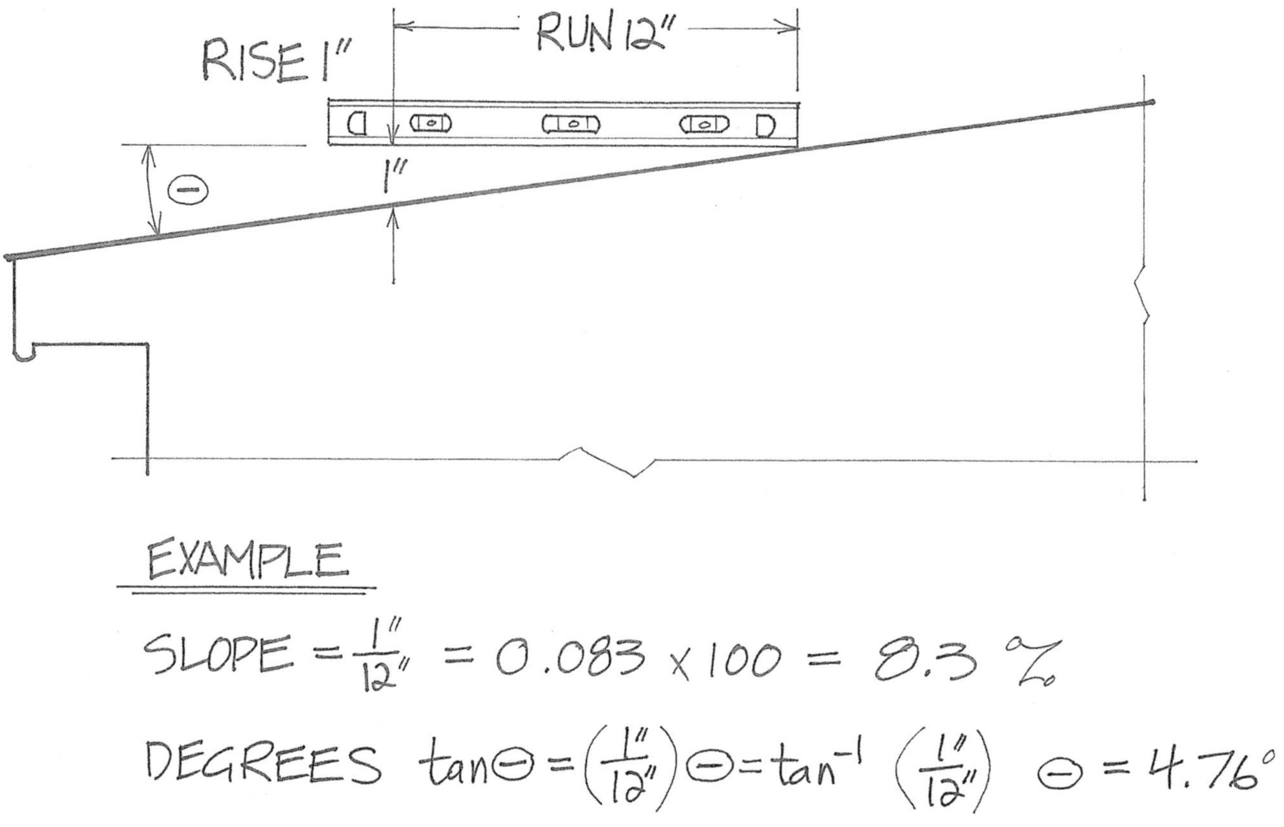

1 inch (2.54 cm) above the 12-inch (30 cm) mark, then the pitch is 1:12 or 8.3%.

Figure 3.13 is a table of common slopes and units, along with notes on necessary precautions to consider for the various slopes.

The rise will be very minimal on roofs that appear to be flat. Be sure that you are

EXAMPLE

$$\text{SLOPE} = \frac{1''}{12''} = 0.083 \times 100 = 8.3\%$$

$$\text{DEGREES} \quad \tan\Theta = \left(\frac{1''}{12''}\right) \quad \Theta = \tan^{-1}\left(\frac{1''}{12''}\right) \quad \Theta = 4.76°$$

Figure 3.13: Roof Slopes with Degree and Pitch Equivalents and Necessary Precautions

Percent Slope	Degree	Roof Pitch Fall Per Foot	Necessary Considerations [12,13]
2%	1.1°	0.24:12	Minimum slope needed for natural drainage
4%	2.3°	0.48:12	Install adequate drainage
5%	2.9°	0.6:12	The drainage layer may be replaced with a water-retention layer
10%	5.7°	1.2:12	Use anti-erosion netting; use root-resistant membranes
17.6%	10°	2:12	Slope restraints required
25%	14°	3:12	Ensure growing media components are not easily washed away; promote quick plant growth to bond growing media
33%	18°	4:12	Consider modules or pre-vegetated mat systems. Consult with an engineer
40%	21.8°	4.8:12	Consult with an engineer for reinforced systems, otherwise not recommended to plant on such a steep slope

not touching any seam points because that can make a big difference in a calculation on a flat roof.

While it is often said that green roofs are installed on flat roofs, flat roofs are never usually completely flat. Almost all flat roofs will have a slope of at least 2%. This minimum slope is intentional; it allows water to exit the roof rather than pool. Greening a roof with no slope can be a challenge because pooling water can keep the media saturated too long, rotting plant roots and keeping excess weight on the structure. If you are working with an existing roof that is absolutely flat, talk with a roofing company about installing sloped insulation or crickets to resolve this problem.

Slopes affect many aspects of the build, including drainage, irrigation, planting, erosion, installation safety, and maintenance.

Drainage

As the slope increases, water will run off more rapidly. Once a slope reaches 5%, a drainage layer is unnecessary; instead, a water retention layer becomes more valuable. A high slope may require more freely draining aggregate to be installed near the base of the slope to allow unimpeded water movement during storm events.

Irrigation

Sloped roofs tend to have different zones of wetness; there will be drier soils near the

In this project, a green roof was installed on a slope of 37.5%. The slope restraint system couldn't be tied into the structure, and the green roof materials experienced some shifting as it settled over the first couple of years; this resulted in a chimney support screw coming loose that let water in.

Fig. 3.14: *A steep slope install.*
Credit: Restoration Gardens, Inc.

top and saturated soils near the bottom. One option to solve this would be to install drip irrigation for just the top section of the roof.

Planting

Because water will be more available in some areas of the roof than others, planting strategies should take these microclimates into consideration. One strategy for planting on sloped roofs is to plant drought-tolerant plants near the top ridge and plants that require more water near the eaves.

In addition to dictating moisture availability, the slope may also affect sun exposure. A slope that faces west will have limited morning sun and intense afternoon sun; a slope that faces north may have less sun exposure. This will affect the plants that you choose for the roof.

Wind Erosion

Wind often scours the tops of sloped roofs. In wind-prone environments, you may want to include wind erosion blankets. These can hold the substrate in place until the plant roots are mature enough to hold the soil in place.

Installation Safety

All green roof installations should be done safely. In sloped-roof applications, this is harder to execute. These roofs typically do not have railings or parapets and can be more of a challenge to walk on safely. You should plan *in advance* your installation safety measures. More on this will be given in Chapter 4.

Maintenance

Plantings that require regular maintenance, especially with equipment such as mowers, should be avoided (especially on slopes) when the homeowner is not comfortable performing these tasks or if walking on the slope has the potential to cause further material slippage. It is best to use low-maintenance, drought-tolerant species on these roofs and only execute maintenance when necessary.

Some small green roofs may come together quickly, while others will require that you carefully consider all these various elements. Take your time and do not rush the planning process. It will be very difficult to make changes once the roof has been completed.

Chapter 4

Roof Access and Safety

You have now determined that your roof can handle the loading, and you are aware of your slope. It is now time to determine how to move your project forward. The first question that gets most people scratching their heads is *how do we get this stuff onto the roof?*

Access to your roof is about more than just how you get on your roof—although that is an important question. Keep in mind that you will need to haul up large, bulky material, get water up onto the roof, and visit your roof at least twice a year for ongoing maintenance. How are you going to get up there? A garden shed roof may only require a 12-foot ladder. A second-story roof might be accessible through a bedroom window, a patio door, or a 32-foot ladder. The higher you get, the more challenging the project becomes.

Safety cannot be stressed enough when building a green roof. In this section, I will address several of the known risks; however, all sites are unique, and there are always unknowns that arise on each site. It is important to compose a list of risks and how they can be approached safely. For example, a complicated site with limited access or the necessity of taking materials up through the house presents certain risks. In these cases, modular systems might be better, safer ways to go. On the other hand, easy-to-access roofs may make it more manageable to build in place.

In this section, I will go through a few ways to access the roof and some considerations for safety.

Access to the Roof
Ladders

Often, people use ladders that are not appropriately sized for the height of the roof. Your ladder needs to be tall enough to not only access the roof, but to be placed a safe distance from the wall at ground level. The safe distance is a 4:1 ratio of roof height to distance out from the wall. To determine the appropriate placement, measure the height from the ground to the roof, and divide that by 4; this tells you how far away from the wall to place the feet of the ladder. The ladder itself should be 3 feet taller than your total height measurement. Refer to Figure 4.1 and 4.2 to help you find the correct ladder.

Ladder Considerations

- Ladders may require more than one person to set up.
- Ladders require solid, level ground and should be tied off securely, ideally to a permanent tie-off point on the roof.
- Ladders may make hauling materials up onto the roof awkward.

28 *Essential* GREEN ROOF CONSTRUCTION

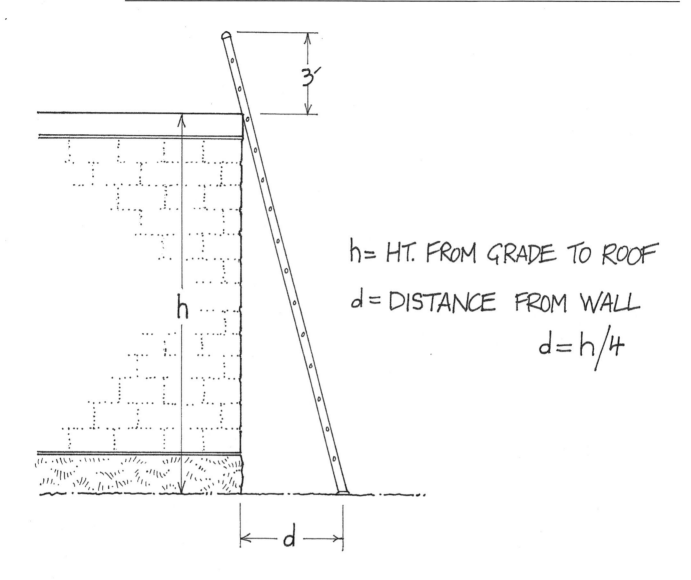

Fig. 4.1 *How to accurately distance the extension ladder feet from the base of the wall.*[1]

Figure 4.2: How High Extension Ladders Can Reach Based on Safe Set-ups*

Extension Ladder Height	Maximum Reach of Extension Ladder	Distance of Base Placement from Wall	Highest Point Ladder Will Touch
16 foot	15 feet	4 feet	9 feet max
20 foot	19 feet	5 feet	9 to 13 feet
24 foot	23 feet	6 feet	13 to 17 feet
28 foot	27 feet	7 feet	17 to 21 feet
32 foot	31 feet	8 feet	21 to 25 feet
36 foot	34 feet	9 feet	25 to 28 feet
40 foot	37 feet	10 feet	28 to 31 feet

*Chart information provided by Sunset Ladders

Fig. 4.3: *While this ladder is extended an appropriate 3 feet above the roof, it is only secured to a post. A better set up would have two tie-off points so the ladder cannot pivot or tip. In addition, the power cord under the ladder should be moved away from the ladder and secured so it does not accidentally shift the ladder if someone were to pull it.*
Photo Credit: Restoration Gardens, Inc.

Window/Roof Hatch Access Considerations

- Ensure there is a safe landing both inside and outside of the window.
- Some bulkier materials may be challenging to get through the window.
- If bringing materials through the house, be prepared for the trail of dirt this will inevitably leave behind.
- For roof hatches, you must be able to tie off the ladder and secure the feet at the bottom to prevent the ladder from shifting while you are on the roof.

Scaffolding Considerations

- Scaffolding requires level ground, railings, assembly, and secure tie-offs.
- As this is a temporary set up for installation, you will require a separate plan for future maintenance of the green roof.

Mobile Elevated Work Platforms

- These include scissor lifts, boom lifts, bucket trucks, and other material-handling equipment.
- Requires a licensed operator.
- Requires level ground.
- Not recommended to use for human access on and off the roof unless fall protection measures are in place.
- As this is a temporary set up for installation, you will require a separate plan for future maintenance of the green roof.

Maintenance and Repair Access

Unless you do not mind uninvited trees, invasive plants, and the occasional "brownout" on your roof, there is no such thing as a maintenance-free green roof. Once the roof is done, you will need access to the roof for maintenance. At the very least, you should be visiting the roof twice a year to ensure that the plants are surviving; you'll need to remove large taproot weeds, and check that weather has not damaged or altered your roof components. Depending on the type of green roof, maintenance may require just a body and a bucket for weeds. Because less equipment is involved, you may find that squeezing through a window or braving a ladder is less of a challenge for maintenance than it was when you were doing the installation.

You will save yourself a lot of late-night worrying or onsite frustration if you take the time to plan your job with safe access for everyone at every stage.

Install Logistics

Once you've established how you can get on the roof, consider how the materials will get to the site.

- Materials purchased in bulk will need a dump site and a clear route for hoisting up to the roof. If there is limited room on site, consider purchasing bagged material.
- If purchasing separate growing media components for mixing, you will need more room in your staging area.
- Ensure you have access to water on the site. Plants will need to be watered upon arrival and after installation.

Fig. 4.4: *There was no water source available at the time of this planting. Water was not available until after the plants had suffered, resulting in die-back and the need to re-plant the roof.*
PHOTO CREDIT: RESTORATION GARDENS, INC.

Rooftop Safety

Working at heights can be extremely dangerous. Slips, trips, and falls are very common on job sites. While homeowners working on their own homes are not subject to labor laws, if you hire someone to help you out, you are legally responsible for their safety.

The Ontario Occupational Health and Safety Act requires sites that expose anyone to a 10-foot (3-meter) fall to have fall prevention strategies in place. These may include safety railings or safety harness systems.[2] If you cannot rent temporary railings to clamp onto parapet walls, you can construct your own temporary rails from lumber.

You can also set up free-standing fences or bump lines. Normally, labor laws require them to be placed 6 feet (1.8 meter) in from the roof edge. While this may impede your ability to efficiently roll out material onto the roof, it will allow you to complete the job safely. If you are building a new roof, instruct the engineer/architect to design a tie-off for fall protection. Refer to your local building code or Ministry of Labor for safe practices.

Here are some basic recommendations:

- Never work alone when building a green roof. Working with at least one other person allows you to discuss and address any install risks that present themselves so you will have safe ways to get materials onto the roof.
- Keep all pathways clear of debris and keep all cords and hoses out of heavily trafficked pathways.
- Weigh down seams with visible markers so you are less prone to trips.
- Arrange building materials in strategic places to prevent them from being trip hazards or blowing off the roof.

Fig. 4.5: *Temporary safety rails are required when the fall is greater than 10 feet.*
PHOTO CREDIT: RESTORATION GARDENS, INC.

> *NOTE:* There are lots of courses available online or in-person on ladder safety, scaffolding set up, lifting equipment operation, and working at heights. For anything you are not comfortable with, consider taking one of these short courses.

- Do not leave the worksite until all debris has been cleared. Sweep up all sawdust, bent nails, etc.; a quick gust of wind can easily blow debris off the roof or into your eyes.
- Do not work in poor weather conditions, including extreme heat, rain, or high winds.
- Never throw anything off a roof.
- Make sure any roof openings have a strong cover that is secured in place and marked as a roof opening. Ideally, all these openings should be completed, waterproofed, and flashed prior to the installation of the green roof.

It may be a challenge to find ways to create a safe roof environment while also allowing for an efficient install. Use your best judgment to determine the safest way to work on the roof.

Site Risks

Take a walk around your site and identify temporary or permanent hazards.

Overhead Wires and Cables

Look for overhead wires and cables that may interfere with material lifts or your ability to work on the roof. Never execute lifts near overhead wires.

Weather

Check the weather ahead of time. Only install during agreeable weather. Installing during strong winds or in rain is not recommended. Wind can blow materials either completely off the roof or around the roof, which can cause you to have the type of knee-jerk reactions that lead to injuries. Rain can make for slick footing on the roof or ground, and it can make it impossible to safely climb a ladder. In hot weather, try to plan an early start or shorter days to avoid heat exhaustion. Working with some roofing materials can make the heat seem even more intense.

It is better to have a stretch of good weather so that you can complete the job without having to take extra precautions to weigh down materials, leave the site unattended for long periods of time, or—the worst—rush through the job without enjoying yourself!

Skill Required

Given a relatively small and uncomplicated roof, I have no doubt that most people can build their own green roof. Building a simple green roof is not tricky. If you are comfortable working at heights, can use a measuring tape and some simple tools (box cutter, hammer, circular saw), and can lift at least 50 pounds (23 kg), you can build a green roof.

Safe Operation of Tools

Operating tools safely can simply mean taking the time to use them right and using the right tools for the job. Proper technique can help ensure your safety, so if you are not familiar with a specific tool, ask someone to help or teach you. Do the job safely, even if it means slowing down.

- Power tools should not be used without proper instruction and personal protective gear, such as protective eyewear.
- Never lower or carry a power tool by its cord.

- Use caution around electricity and plumbing. Reserve skilled jobs for the appropriate tradesperson.
- Remember that most power tool accidents happen after the material has been cut and the tool is in "wind down."
- When cutting materials, always use a sharp blade; they are much safer than a dull blade.
- Do not use tools for an unintended purpose.
- Wear work gloves and work boots to protect yourself from sharp objects.

When to Include Tradespeople

There may be times when you have tradespeople working on aspects of the building that affect the roof, such as running plumbing vents through the roof or installing skylights. Make sure all tradespeople are aware of the location of the green roof and that care is taken to protect the integrity of the membrane. Discuss with the tradespeople any future access they may need to the roof for maintenance (visit frequency, repair logistics, etc.). For example, if you have an A/C unit on the roof, you will want to create a path to the unit, place the unit on concrete pavers, and ensure plants do not grow within a minimum of 12 inches around the unit so tradespeople can do their work without damaging the plants. Ideally, all tradespeople should have their work completed before you start.

Some of you may want to hire an experienced roofer to install the membrane for you. That is O.K.; in fact, we do this on many of our urban residential projects. However, it is crucial that installers are fully aware that the membrane will be used in a green roof assembly; include in the contract any implications this may have.

There are some big differences between having professionals build you a green roof and doing it yourself:

Experience: This can account for a lot. As you have already read, many risks are involved in an installation. How many and what kind depends on your site and the size of the job. A professional will already know how to handle your roof, while you are still on a steep learning curve. If you are working on a very high roof, being uncomfortable with tools can make the job seem even harder and add stress to what should be an enjoyable project. If this is the case, at the very least, recruit a friend with building experience to help. They can help you create a safe site and manage the more challenging tasks.

Warranties: There is something to be said for the ease of mind that comes with professionals offering warranties in the event that something happens with your green roof. A professional roofer can offer membrane warranties even if a green roof is installed above, but be sure to confirm this beforehand. On the other hand, for a green roof over a garden shed, a warranty may seem irrelevant, and you can have the fun of building it yourself.

Material Flexibility: This is in a DIYer's favor! The benefit of doing it yourself is that you can use materials that are available locally or that you may already have on hand—if they provide the same function as those used by professionals. Professionals will have a series of products they use based on proven success and/or manufacturer warranties.

Planning a green roof, like most things in life, gets easier the more times you do it. Your first project will go smoothly only if you take the time to plan it right. I suggest you read this book all the way through, and then come back to the accessibility and install considerations listed in this chapter. Once you know the specific materials you want to work with (and amounts/sizes), you can better answer how you can get them on your roof. Above all, be safe. Work smart and be aware of all the hazards. Assess hazards during the design phase as well as every morning before you get to work.

Chapter 5

Site and Design Factors

UNDERSTANDING MUNICIPAL regulations, structural allowances, and access and safety conditions can determine whether or not a green roof is feasible. It can also guide you toward the appropriate green roof class and type of system. Once you have that information, you can begin tackling the design. Remembering your structural details and the site safety notes from the previous chapters, create one or two drawings, including a roof plan and a site plan. A site plan will ensure you consider factors that will affect your material and plant selection.

Rooftop Measurements

Below are the types of measurements that are required for building a green roof. I have given an example of both a new roof and an existing roof.

The first measurements we need is the area (L × W) of your roof deck. For new roofs, it will likely stretch from all the edges of your roof deck. Other measurements needed are the slope direction and percentage, and the height from grade. You will also need to decide how the roof will drain.

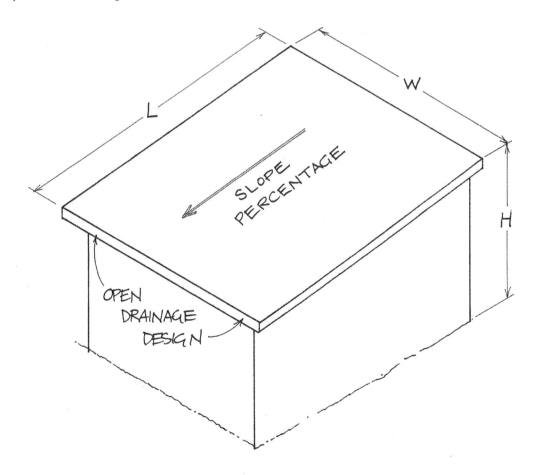

Fig. 5.1: *Measurements required for a new build.*

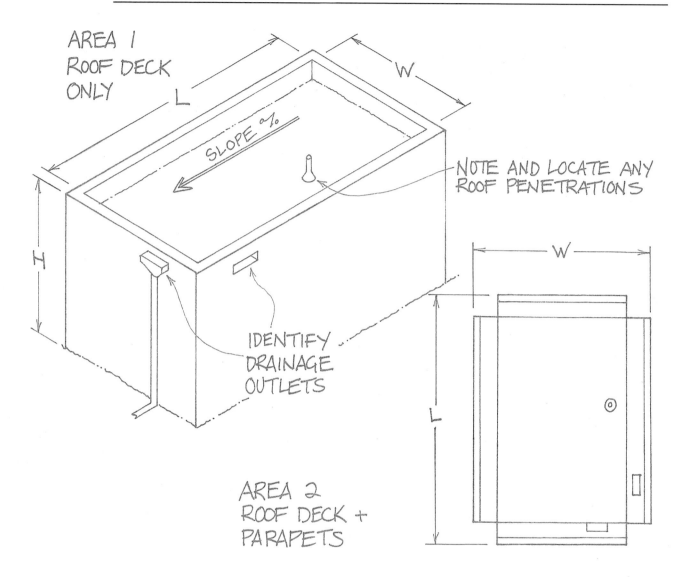

Fig. 5.2: *Measurements required for an existing roof.*

For existing roofs, you will need the total area. For existing roofs with parapets, you will need to know the area within the parapets as well as the area including parapets. This is because some materials only go on the roof deck, while others, such as the membrane, root barrier, and filter cloth, go up and over the parapets. Keep in mind that due to the slope of the roof, your parapet heights may change from one end to the other. When calculating the measurements with the parapets, think of it as measuring a flattened box. If the green roof abuts a wall, be sure to account for the vertical distance as required by your local code for waterproofing.

Collect measurements for the area, height from grade, and the sizes (and locations) of any mechanical systems. Note the location of drains. Find the roof slope direction and calculate its percentage (as discussed in Chapter 3).

Once you have this information sketched out in your roof plan, you will want to include your structural loading capacity and make notes about where you will access the roof (during install and for maintenance), where you will stage material, and where your water source is.

Rooftop Mechanicals or Penetrations

Indicate on the roof section where any mechanical systems or penetrations are located on the roof. These might include air conditioning units, skylights, and plumbing vents. Vegetation-free zones should surround all of these to prevent vulnerable areas from being overgrown with vegetation. These areas should be completed and flashed prior to the installation of your green roof. If your municipality requires you to follow a standard or guideline, they will most likely include specifications for vegetation-free zones, but even if not required, keeping these areas clear is important in terms of increasing drainage, creating fire breaks, and providing easy access in the event of maintenance or repair.

Site Environment
Sun Exposure

To guide you toward appropriate plants, study the roof for your sunlight exposure. In general, plants are classified according to the hours of sunlight per day that they require.

- **Full Sun:** Direct summer sun for six hours or more per day
- **Partial Sun:** Areas that receive two to six hours of sun per day
- **Partial Shade:** Areas that receive two hours of direct sun each day or are shaded for at least half the day. Can also refer to areas that receive cooler morning sun with little or no hot afternoon sun.
- **Full Shade:** Less than an hour of direct sunlight each day or dappled light through tree canopy for most of the day.

On your plan, identify the following:

- The direction that your roof faces.
- Any large objects or structures that may affect sunlight, such as trees or adjoining houses. Large south-facing windows that line the green roof will reflect a significant amount of light and heat to the plants.

Here are some helpful tips:

- Indicate *full sun* if it is a flat roof without any surrounding structures or trees.
- Indicate *partial sun* if your roof is sloped and facing the north, as you can rely on morning and afternoon sun.
- Indicate *partial shade* if you have a roof that is flat but has large trees on the south and west, as you'll receive cooler morning sun with less midday sun.
- Perhaps your roof will only receive afternoon sun; this would be *partial sun*, but keep in mind this can lead to extreme fluctuations in temperature, as morning sun is cooler and afternoon sun can be quite hot.
- Indicate *shade–partial sun* if your roof is on the north side of your house with structures blocking the sun from the south and east.

Keep in mind that you may have different exposure levels for different sections of the roof. Try to identify them all as best as you can. This will allow you to increase plant diversity on the site and have a better chance at achieving full coverage. Too-deeply shaded areas may be best reserved for moss (see Chapter 6) or decorative stone/pavers. Such areas can be ideal access points, allowing you to step onto the roof without damaging plants. Photos 6 and 7 in the Color Section show one roof with two very different exposures.

Climate

The two most important factors in the climate of an area are *temperature* and *precipitation*. For Canadians, Agriculture Canada created a Plant Hardiness Zone map to identify the different regions in which trees, shrubs, and flowers can survive. Differing from the United States Department of Agriculture (USDA) version, which divides regions by the average annual minimum winter temperature, the Canadian one takes into consideration minimum winter temperatures, length of the frost-free period, summer rainfall, maximum temperatures, snow cover, January rainfall, and maximum wind speed. The Canadian map has 9 major zones; the harshest climate is 0 and the mildest is 9, and the zones are further subdivided into 19 subzones. The USDA version has 13 zones. The two numbering systems do not match between countries. In most cases, the Canadian map zones are one higher than the USDA zones. For example, a US zone 5 is a Canadian zone 6. See Appendix C for links to hardiness zone maps.

There are many maps available that give expected annual precipitation. With our changing climate, this is sure to change; you

Fig. 5.3: *This roof was planted in full shade with full overhead cover. Even though the owners tried to accommodate for the lack of rain with irrigation, the plants they chose did not survive.*
PHOTO CREDIT: RESTORATION GARDENS, INC.

Fig. 5.4: *This large gable roof will add a significant amount of water to the adjoining porch roof designed for plants.* PHOTO CREDIT: RESTORATION GARDENS, INC.

Fig. 5.5: *This green roof utilizes stones and thirsty plants to mitigate surface erosion and take advantage of the water draining from an upper roof.* Photo Credit: Restoration Gardens, Inc.

may have already noticed a greater frequency of 100-year flooding in your area and/or longer, more intense heat waves.

Include your hardiness zone and your annual precipitation on your design plan. These factors will dictate which optional layers you might want, and they will guide your plant choices. For example, if you live in an arid region, you may want to include water retention fabrics or design your growing media to include components that have greater water-holding capabilities. In areas with heavier rainfall, you may want the roof to drain quickly, and you will likely be able to increase the range of plants beyond those that are drought tolerant. Make special note of objects that may interfere with precipitation; for example, a large, dense tree canopy over the roof might reduce the rainfall that reaches a portion of the vegetation. Or, an overhead shed roof that drains onto your green roof will increase the amount of water reaching that part of the roof; this could be a spot for thirsty plants.

Wind

Wind is a beautiful and necessary element of nature, aiding in pollination and airflow. However, it can also affect your install and wreak havoc on your green roof. Wind can affect your hoisting schedule or blow loose items off the roof during the install. As wind travels up your building and across the open roof, it can create a negative pressure, leading to uplift (or the ballooning) of the layered material. This is why membranes are typically fully adhered, mechanically fastened, or covered with ballast. In addition, as wind gets redirected around parapets and abutting wall structures, it can swirl around in what is

Fig. 5.6: *Wind scour exposed underlying layers on this site before the plants could provide coverage.*

PHOTO CREDIT: RESTORATION GARDENS, INC.

called a wind vortex. Wind vortices can lead to displacement of your growing media, exposing roots or layers to further damage. You must design the system so that the weight of your green roof layers (specifically, the growing media) is sufficient to hold down the assembly against these forces.

Wind is site-specific. It will vary by the contours of your land, large vegetation screens, and the design features of your roof, such as building height and parapet structures. Wind maps indicate regional wind zones, and these can be useful in determining the wind speeds your site may endure and what provisions, if any, will be necessary to determine your wind uplift pressures; this will allow you to design an appropriate system. For roofs exposed to strong winds or on large residential projects, it is advisable to work with an engineer. Where wind speeds periodically exceed 140 mph (62 m/s), it is recommended that you avoid building a green roof altogether. These high wind speeds typically affect areas along the southeastern coasts of the US. If you live in a high-wind region and still want a green roof, consult with an engineer to design a suitable system.[1]

Solutions to mitigate the damage by wind can include perimeter ballasting or covering the vegetative surface with temporary or permanent netting. These options are discussed in Chapter 7.

On the sketch of your site, make a note of your wind exposure. Keep in mind the following considerations:

- High flat-roof surfaces without surrounding shelter should be considered at risk for wind scour or uplift forces.
- Sheltered urban or low roofs are normally low risk for wind damage.
- Roofs with connecting walls or taller parapets may be at risk for wind vortex pressures.
- Regions prone to tornadoes or heavy winds will need engineered wind-protection measures. Building a green roof in these regions may be ill-advised.

Figure 5.7 offers an example of important information to note on site plans and preliminary designs.

Never underestimate the benefits of careful planning. Looking over your roof space on paper multiple times can save you many headaches later. Work through this chapter to ensure that you have at least considered all the elements. Once you've read through the plant list and material choices that are included in the next chapters, you'll be able to sketch out a few different ways you can build the roof to ensure you create a roof that will be a success in your region, on your structure, and within your maintenance expectations.

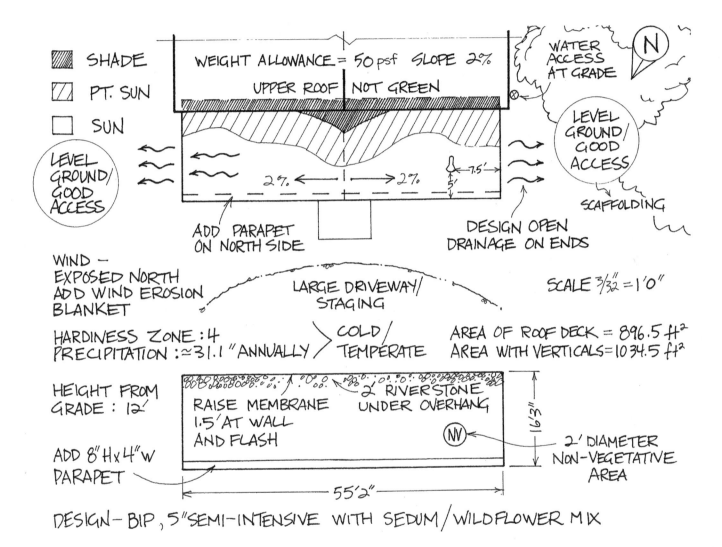

Fig. 5.7: An example of site planning and roof notes necessary to design a green roof.

Chapter 6

Plants

Choosing plants for your green roof is an exciting part of the process. While the plant list is more limited than for at-grade gardens, the choices are still plentiful. With many beautiful species to choose from, you can let your rooftop features help determine the best plants for your roof.

The Green Roof Growing Environment

The green roof environment differs considerably from gardens at grade. Because a roof has weight restrictions, drainage requirements, and accessibility issues, growing media tends to consist of shallow, lightweight aggregates with low organic content. As a result, extensive and semi-intensive green roofs have growing environments that are dry, shallow, and nutrient poor, with intense heat and higher exposure to wind. Even though a substrate can be amended with more organic matter, green roofs are not the location for beloved roses or peonies. The place for these is on the ground in deep, rich soil in sheltered locations. When creating a green roof garden, we strive to create sustainable landscapes that require fewer resources. As a result, the harsh rooftop environments are more reminiscent of alvars, alpine settings, prairies, meadows, or even roadsides.[1]

Alvars are open habitats that occur on shallow limestone or dolostone bedrock with very little soil. (See Photo 8 in the Color Section.) They often suffer seasonal flooding as well as drought. Alpine communities also suffer from drought, as well as strong winds, low winter temperatures, low nutrient availability, and short growing seasons. The plants that grow in these environments have shallow roots that spread out in search of nutrients and provide anchorage against the wind. Although some plants in these communities rely on deep taproots, they survive only because they can find deep crevices that collect debris and moisture, thus ensuring their survival. Green roofs do not offer these opportunities, so plants with these root systems should be avoided. The exception to this rule is in a food production roof, where vegetables such as beetroots, radishes, and even carrots can be successfully grown.

"Prairie" or "meadow" systems are often promoted by green roof companies. These systems can be designed for semi-intensive green roofs that have a substrate relatively high in organic matter—to sustain the growth of many herbaceous perennials. Green roofs *can* be designed for some prairie and meadow species, as both ecosystems are home to some species that are drought and stress-tolerant and require very little nutrients. However, not all prairie and meadow species will grow on a green roof. Some species require the deep soil profiles that prairies offer. Some meadow species require the richer, moist soils of a natural meadow. Both of these ecosystems also have natural microbial communities that some species depend on; these are a challenge to replicate on a roof.

The plants you see growing on roadsides are often well-suited for green roofs. Roadsides are often gravel based and are exposed to strong winds. Plants that can survive there are opportunistic and can withstand harsh environments that offer few nutrients. (Note that *roadsides* are not the same thing as *ditches,* which can contain quite a lot of moisture and have wind protection.) So, you might want to take notice of what is growing by the road when you are out driving.

Plant Types

Due to the harsh growing environment of a green roof, what you need to plant are low-growing, shallow-rooted perennials that are tolerant of a wide range of stresses. Focus on stress tolerance; plants need to be able to survive drought, wind, insects, disease, and temperature fluctuations. Species that are short-lived or need regular care and fertilization are sure to disappoint. The aim is to reduce the amount of planting and maintenance that is required on the roof.

Native Plants

Native plants provide valuable food and nesting habitats for native species of birds and insects. If left unmown, seed heads offer much-needed food sources for birds in the winter as well as winter interest in the garden. Natural ecological communities, such as alvars and tallgrass prairies, are disappearing, and green roofs can offer protected places for some of their rare species of plants. Just because a plant is native, however, does not make it suitable for a green roof. Suitable species should display the characteristics noted above.

Succulents

Succulents are plants with thick, fleshy leaves; they do extremely well in the green roof environment. They can survive drought and wind conditions through a metabolic process called *crassulacean acid metabolism* (CAM). This conservation process means they open their pores at night rather than during the day—the opposite of what most plants do. This reduces the amount of transpiration that occurs during the day. If, however, you are planning a green roof for maximum stormwater retention or for keeping the roof surface cooler, succulents may not be the best choice because they store water in their leaves and transpire only at night. This results in less water uptake and less transpiration during the warmer daytime temperatures. In addition, these plants thrive in a free-draining aggregate media with little organic matter. This type of growing media retains less water and can hold a bit more heat.

Sedum is one type of widely available succulent, and it is a popular choice on extensive green roofs. While there are some North American native species of *Sedum*

Fig. 6.1: *Sedums have fleshy leaves that store water, making them extremely valuable in green roof planting plans. Here,* Sedum *'SunSparkler Lime Zinger.'* Photo Credit: Restoration Gardens, Inc.

(i.e., *Sedum oreganum*—native to Oregon and *Sedum ternatum*—native to Missouri), there are many nonnative species that do well on green roofs. Their success is due to their ability to provide rapid coverage with low maintenance and low nutrient needs. Rapid coverage reduces opportunities for unwanted plants to blow in. You will never run out of options with *Sedum*, as there are hundreds of varieties to choose from.

Other succulents that perform well on a green roof include *Delosperma, Jovibarba, Opuntia, Sempervivum,* and *Talinum,* some of which have species native to North America.

Lifecycles

Almost all of the plants that are recommended in this book are perennials. Perennials grow, flower, and set seed in one or more growing seasons, and they do not die after setting seed. This is in contrast to annuals, which grow, flower, set seed, and die in one growing season. Annuals are not recommended in large quantities because they require more water than perennials do to reduce wilting; also, if they do not self-seed, you have to fill in the gaps they leave behind. A few biennials are recommended. Biennials grow vegetation the first growing season, and then they flower, set seed, and die the second growing season.

Self-Seeding Plants

Some plants will just continue to grow larger in the same spot you planted them, while others will multiply through their ability to self-seed. Self-seeding occurs after the flower blooms. The seeds will develop and be dispersed, then go through a dormancy period, and then sprout new plants the following spring. While self-seeders can provide a cost-effective way to fill in a garden, note that some gardeners find them to be a nuisance because they can spread too far and too easily. This may be the case if you are planting a roof in a dense urban setting. Your neighbors may not appreciate unexpected new plants in their rose or vegetable garden. If your roof is fairly sheltered or remote, however, self-seeders can be excellent choices. *Allium schoenoprasum* (common chive) is one self-seeder that is often included on green roofs. See examples in Photos 10, 17, and 19 in the Color Section.

Fig. 6.2: Rudbeckia hirta *(black-eyed Susan) is a favorite biennial that fills in gaps if flower heads are left alone and allowed to drop seed over the winter.* Photo Credit: Restoration Gardens, Inc.

Recommended Plants

In Figure 6.3, I list over 100 species that are suited to the green roof environment. But even this list is not comprehensive; there are definitely more! You will find others in helpful books like *Green Roof Plants* by Edmund and Lucie Snodgrass; some other good books to consult are listed in the Bibliography.

Plants in my list are divided by their life forms, such as broadleaf herbaceous, succulents, ferns, and grasses; recommended minimum depths of growing media are given for each.

In the recommended plant list, I have identified whether or not a species is native, with "native" referring to plants native to North America. If you are looking to incorporate plants that are native to your region, you can take this plant list to your local nursery for further classification or speak with a local ecology group.

I have tried, as much as possible, to include plants listed in all growing zones throughout North America. Most information available identifies USDA growing zones, so this is the system identified in the list. Remember that Canadian growing zones are typically one higher than the US zones. Also, remember that there are always anomalies when it comes to hardiness zones. Due to onsite microclimates created by features unique to your roof or due to changes from global warming, you may be able to grow species outside your hardiness zone. If species are available at nurseries in your region, they are likely suitable.

Sun exposure is listed and fairly straightforward. A species identified as "Sun–Shade" will be tolerant of a variety of exposures on the roof. Shade species may benefit from supplemental irrigation because rain may be blocked by the shade-casting structure or tree; as well, many shade species naturally grow in soils that are moist—as less evaporation occurs within low-lit areas. But note that supplemental irrigation does *not* mean keeping the area saturated.

Plants that are self-seeding are identified, so if you are looking to have a meadow effect, these are the plants to look for. Shorter self-seeding plants may form a mass planting similar to a ground cover. Some self-seeders are not as competitive or are short-lived and therefore should only be planted as an accent on a green roof. Groundcovers spread out from the spot they were planted and provide a rapid, reliable, and cost-effective cover over the roof. Many *Sedum* species can grow 6 to 10 inches horizontally in the first year.

Finally, I have listed flower color and heights. Plants that grow under 12 inches in height are listed as *low*, 12–24 inches are *medium*, while the *tall* plants are 24 inches or higher. Keep in mind that plants may appear stunted on a green roof in comparison to those grown in richer, deeper, at-grade soils.

Once you start to understand the growing conditions of a green roof, you will likely be able to identify additional species. If you have an easily accessible roof and are very hands-on, feel free to experiment and do some trials with your favorite ones.

Figure 6.3: Recommended Plants for Simple Extensive or Semi-Intensive Green Roofs

Latin Name	Common Name	Native to N. America	USDA Zone	Sun	Self-Seeding	Garden Value**	Flower Color	Height
Broadleaf Herbaceous (minimum depth of 4 inches)								
Allium acuminatum	Tapertip onion	Yes	7	Sun–Shade	Yes	Accent	Purple	Low
Allium cernuum	Nodding wild onion	Yes	4	Full/Partial Sun	Yes	Accent, Meadow	Purple/Pink	Medium
Allium schoenoprasum	Common/wild chive		4	Full/Partial Sun	Yes	Meadow	Pink	Low
Allium senescens subsp. *Montanum* var. *glaucum*	German/mountain garlic		5	Full/Partial Sun	Yes	Accent	Pink	Low
Allium tuberosum	Garlic chives		7	Full/Partial Sun	Yes	Accent	White	Medium
*Aquilegia canadensis**	Wild columbine	Yes	3	Full/Partial Shade	Yes	Accent	Red	Medium
*Coreopsis lanceolata**	Lanceleaf coreopsis	Yes	4–9	Full Sun	Yes	Meadow	Yellow	Tall
*Dianthus deltoides**	Maiden pink		3–8	Full Sun		Evergreen, Accent, Groundcover	Pink	Low
Fragaria virginiana	Wild strawberry	Yes	5–9	Full/Partial Sun		Groundcover, Good for bottom of slope	White	Low
Geum triflorum	Prairie smoke	Yes	3–7	Full Sun Partial Shade	Yes	Meadow	Pink	Low
*Iris pumila**	Dwarf iris		4–9	Full/Partial Sun		Accent	Purple	Low
Liatris aspera	Rough blazing star	Yes	3–8	Full Sun	Yes	Meadow	Purple	Medium
Liatris cylindracea	Cylindrical/dwarf/Ontario blazing star	Yes	4–7	Full Sun		Accent, Meadow	Purple	Medium
Malephora crocea var. *purpureocrocea* 'Tequila Sunrise'	Tequila sunrise ice plant		8	Full Sun		Groundcover	Orange	Low
Orosachys boehmeri	Duncecap		6–10	Full Sun		Groundcover	White	Low
Orostachys malacophyllus	Green duncecap		5–8	Full Sun		Groundcover	Yellow	Low
Othonna capensis	Little pickles		9	Full/Partial Sun		Accent	Yellow	Low
Penstemon hirsutus	Hairy beardtongue	Yes	4	Full/Partial Sun	Yes	Meadow	Pink/Purple	Medium
*Phlox subulata**	Moss/creeping phlox	Yes	3–9	Full Sun		Groundcover	Pink	Low
Potentilla simplex	Common cinquefoil	Yes	3–7	Full Sun		Groundcover	Yellow	Medium
Ranunculus fascicularis	Early buttercup	Yes	3–7	Sun–Shade	Yes	Meadow	Yellow	Low

Figure 6.3: Continued

Latin Name	Common Name	Native to N. America	USDA Zone	Sun	Self-Seeding	Garden Value**	Flower Color	Height
*Sisyrinchium montanum**	Common blue-eyed grass	Yes	3–9	Sun–Shade		Accent	Blue	Medium
*Thymus citriodorus**	Lemon thyme		5–8	Full Sun		Groundcover	Pink	Low
*Thymus serphyllum**	Creeping thyme/mother of thyme		4–8	Full Sun		Groundcover	Purple	Low
Broadleaf Herbaceous (minimum depth of 6 inches)								
Achillea millefolium	Yarrow	Yes	2	Full Sun	Yes	Meadow	White	Medium
Agastache foeniculum	Anise/lavender hyssop	Yes	4–8	Full/Partial Sun	Yes	Meadow, Biennial	Purple	Tall
Armeria martima 'Alba'	White sea thrift		3	Full Sun		Accent	White	Low
Armeria maritima 'Splendens'	Sea thrift		3	Full/Partial Sun		Accent	Pink	Low
Artemisia stelleriana 'Silver Brocade'	Silver brocade artemisia		2–9	Full Sun		Groundcover	Silver/White	Low
Aster alphinus 'Albus'	Alpine aster	Yes	2–9	Full Sun	Yes	Accent, Short-lived	White	Low
Aster alpinus 'Dunkel Schone'	'Dark Beauty' aster	Yes	5–7	Full Sun	Yes	Accent, Biennial	Purple	Low
Campanula rotundifolia	Harebell	Yes	2–7	Full/Partial Sun		Accent	Purple	Low
Cerastium tomentosum	Snow-in-summer		2	Full Sun		Groundcover	White	Low
Dianthus alpinus	Alpine pink		4	Full/Partial Sun		Accent, Groundcover	Pink	Low
Dianthus arenarius	Sand pink		5	Full Sun		Accent, Groundcover	White	Low
Dianthus carthusianorum	Carthusian pink		5–9	Full Sun	Yes	Evergreen, Accent, Groundcover	Pink	Medium
Echinacea pallida	Pale purple coneflower	Yes	3–10	Full/Partial Sun	Yes	Meadow	Purple	Medium
Echinacea purpurea	Eastern purple coneflower	Yes	3–8	Full/Partial Sun	Yes	Meadow	Purple	Medium
Erigeron linearis	Desert yellow fleabane/daisy	Yes	4	Full Sun		Meadow	Yellow	Low
Euphorbia myrsinites	Donkey-tail, myrtle spurge		6–9	Full Sun	Yes	Groundcover	Yellow	Low
*Gentianella quinquefolia**	Stiff gentian	Yes	4–7	Full/Partial Sun	Yes	Meadow	Purple	Medium
Geranium sanguineum	Bloody cranesbill, bloodred geranium		3–9	Full/Partial Sun		Accent	Pink	Low
Hedyotis longifolia	Long-leaved bluets	Yes	3–8	Sun–Shade		Accent	White/Blue	Low

Figure 6.3: Continued

Latin Name	Common Name	Native to N. America	USDA Zone	Sun	Self-Seeding	Garden Value**	Flower Color	Height
Hemerocallis 'Stella De Oro'	'Stella de Oro' daylily		3–10	Full/Partial Sun		Accent	Yellow	Low
Liatris spicata*	Dense blazing star/ gayfeather	Yes	3–9	Full Sun		Accent, Meadow	Purple	Tall
Monarda fistulosa	Wild bergamot	Yes	3–9	Full/Partial Sun	Yes	Meadow	Purple	Tall
Nepeta x faassenii 'Walker's Low'*	Catmint 'Walker's Low'		4–9	Full/Partial Sun		Accent, Meadow	Purple	Tall
Oenothera caespitosa	Tufted evening primrose	Yes	4	Full Sun	Yes	Meadow	White	Low
Oenothera macrocarpa	Missouri evening primrose	Yes	4–7	Full Sun	Yes	Meadow	Yellow	Low
Origanum vulgare	Oregano		5	Full Sun		Accent	Purple	Low
Papaver alpinum (hybrids)	Alpine poppy hybrid		3–9	Full Sun	Yes	Accent, Short-Lived	Multi	Low
Penstemon digitalis	Foxglove beardtongue	Yes	2–8	Full Sun	Yes	Meadow	White	Tall
Penstemon pinifolius	Pineleaf beardtongue	Yes	4–9	Full Sun		Accent, Meadow	Red	Low
Penstemon smallii	Small's beardtongue	Yes	5–8	Full Sun-Partial Shade	Yes	Meadow	Pink	Low
Petrohagia saxifraga	Tunic flower		5–7	Full Sun	Yes	Groundcover	Pink	Low
Pycanthemum tenuifolium*	Hairy mountain-mint		4–8	Full/Partial Sun		Meadow	White	Tall
Rudbeckia hirta*	Black-eyed Susan	Yes	3–7	Full Sun	Yes	Meadow	Yellow	Tall
Salvia officinalis	Common sage		4–8	Full/Partial Sun		Meadow	Purple	Tall
Solidago nemoralis	Gray goldenrod	Yes	3–9	Full Sun	Yes	Meadow	Yellow	Medium
Solidago ptarmicoides	Upland white aster, upland white goldenrod	Yes	3–8	Full Sun	Yes	Meadow	White	Medium
Symphyotrichum laeve	Smooth aster	Yes	3–8	Full Sun	Yes	Meadow	Purple	Tall
Symphyotrichum oblongifolium 'October Skies'	Aromatic aster	Yes	3–8	Full Sun	Yes	Meadow	Purple	Medium
Symphyotrichum oolentangiense	Sky blue aster	Yes	3–8	Full Sun	Yes	Meadow	Blue	Tall
Tradescantia ohioensis	Spiderwort	Yes	4–9	Sun-Shade	Yes	Accent	Blue	Medium–Tall
Verbena simplex	Slender vervain	Yes	7–10	Full Sun	Yes	Meadow	Pink	Medium
Verbena stricta	Hoary vervain	Yes	4–7	Full Sun	Yes	Meadow	Purple	Tall
Veronica whitleyi	Whitley's speedwell		3	Full Sun		Groundcover	Blue	Low

Figure 6.3: Continued

Latin Name	Common Name	Native to N. America	USDA Zone	Sun	Self-Seeding	Garden Value**	Flower Color	Height
Broadleaf Herbaceous (8 inches)								
Asclepias verticillata	Whorled milkweed	Yes	4–9	Full/Partial Sun	Yes	Meadow	White	Tall
Lavandula angustifolia 'Hidcote'	English lavender		5–8	Full Sun		Accent	Purple	Low
Ferns (minimum depth of 4 inches)								
Cheilanthes lanosa*	Hairy lip fern	Yes	5–9	Full Sun-Partial Shade		Evergreen, Accent	Green	Low
Ferns (6–8 inches)								
Dryopteris marginalis	Marginal wood fern	Yes	3–8	Full/Partial Shade		Evergreen, Accent	Green	Medium
Polystichum acrostichoides	Christmas fern	Yes	3–9	Full/Partial Shade		Evergreen, Accent	Green	Medium
Dennstaedtia punctiloba	Hay-scented fern	Yes	3–8	Full/Partial Shade		Evergreen, Groundcover	Green	Medium
Cystopteris bulbifera	Bulblet fern	Yes	3–6	Full/Partial Shade		Evergreen, Accent	Green	Medium
Polypodium virginianum	Rock fern	Yes	3–8	Full/Partial Shade		Evergreen, Groundcover	Green	Low
Succulents (minimum depth of 2 inches)								
Sedum acre			2–9	Full Sun		Evergreen, Groundcover (aggressive spreader)	Yellow	Low
Sedum sexangulare			2–9	Sun–Shade		Evergreen, Groundcover (aggressive spreader)	Yellow	Low
Succulents (minimum depth of 4 inches)								
Delosperma basuticum 'Gold Nugget'	Gold nugget ice plant		6	Full Sun		Evergreen, Groundcover	Yellow	Low
Delosperma cooperi	Ice plant		6–10	Full Sun		Evergreen, Groundcover	Pink	Low
Delosperma dyeri	Dyer's ice plant		5–9	Full Sun		Evergreen, Groundcover	Red	Low
Delosperma nubigenum 'Basutoland'	Yellow ice plant		5–10	Full Sun		Evergreen, Groundcover	Yellow	Low
Jovibarba (3 species available)	The other hens and chicks		2–9	Full Sun		Groundcover	Yellow	Low
Opuntia humifusa	Eastern prickly pear cactus	Yes	4–9	Full Sun		Accent	Yellow	Low
Sedum 'Matrona'			3–9	Full Sun		Accent	Pink	Medium

Figure 6.3: Continued

Latin Name	Common Name	Native to N. America	USDA Zone	Sun	Self-Seeding	Garden Value**	Flower Color	Height
Sedum album			4–8	Full/Partial Sun		Evergreen, Groundcover (aggressive spreader)	White	Low
Sedum cauticola 'Bertram Anderson'			2–9	Full Sun		Accent	Purple/Red	Low
Sedum cauticola 'Lidakense'			2–9	Full Sun		Accent	Dark Pink	Low
Sedum hispanicum			5–10	Full Sun		Evergreen, Groundcover	White	Low
Sedum hybridum 'Immergruchen'			3–9	Full Sun		Groundcover	Yellow	Low
Sedum kamtschaticum			3–8	Full/Partial Sun		Groundcover	Yellow	Low
Sedum kamtschaticum var. *floriferum* 'Weihenstephaner Gold'			2–9	Full/Partial Sun		Groundcover (aggressive spreader)	Yellow	Low
Sedum middendorffianum var. *diffusum*			3–8	Full Sun		Groundcover	Yellow	Low
Sedum montanum			5–9	Full Sun		Evergreen, Groundcover	Yellow	Low
Sedum oreganum		Yes	2–9	Sun–Shade		Evergreen, Groundcover	Yellow	Low
Sedum pachyclados			3–9	Full Sun		Evergreen, Groundcover	Pink	Low
Sedum reflexum 'Blue Spruce'			3–9	Sun–Shade		Evergreen, Groundcover	Yellow	Low
Sedum rupestre 'Angelina'			5–8	Full Sun		Evergreen, Groundcover	Yellow	Low
Sedum spurium			3–9	Full Sun		Evergreen, Groundcover	Pink	Low
Sedum spurium 'Album'			3–9	Sun–Shade		Evergreen, Groundcover	White	Low
Sedum spurium 'John Creech'			2–9	Full Sun		Evergreen, Groundcover	Pink	Low
Sedum spurium 'Summer Glory'			2–9	Sun–Shade		Evergreen, Groundcover (aggressive spreader)	Dark Pink	Low
Sedum spurium 'Voodoo'			2–9	Full Sun		Evergreen, Groundcover	Red/Pink	Low
Sedum ternatum		Yes	4–8	Full/Partial Shade		Evergreen, Groundcover	White	Low

Figure 6.3: Continued

Latin Name	Common Name	Native to N. America	USDA Zone	Sun	Self-Seeding	Garden Value**	Flower Color	Height
Sempervivum spp. (various species)	Hens and chicks			Full Sun		Accent, Groundcover	Various	Low
Talinum calycinum	Large-flowered rock pink	Yes	6–9	Full Sun	Yes	Groundcover	Pink	Low
Talinum parviflorum	Sunbright, small fameflower	Yes	3–8	Full Sun	Yes	Accent, Groundcover	Pink	Low
Talinum teretifolium	Quill fameflower	Yes	6	Full Sun	Yes	Accent, Groundcover	Pink	Low
Succulents (6 inches)								
Sedum 'Autumn Joy'			2–11	Full/Partial Shade		Accent	Pink	Medium
Grasses (minimum depth of 4 inches)								
Bouteloua curtipendula	Side-oats grama	Yes	3–9	Full Sun	Yes	Accent, Meadow	Brown/Purple	Tall
*Bouteloua gracilis**	Blue grama grass	Yes	3–10	Full Sun	Yes	Accent, Meadow	Red/Purple	Medium
*Deschampsia cespitosa**	Tufted hair grass	Yes	2–9	Partial Shade		Meadow	Purple/Golden	Tall
*Panicum virgatum**	Switch grass	Yes	3–9	Full/Partial Sun		Meadow	Pink/Cream	Tall
Schizachyrium scoparium	Little bluestem	Yes	3–9	Full/Partial Sun	Yes	Meadow	Purple/Bronze	Tall
Grasses (6 inches)								
Carex flacca	Blue zinger sedge		4–9	Partial Sun		Evergreen, Accent, Groundcover	Green	Low
*Carex flaccosperma**	Blue wood/thinfruit sedge	Yes	5–8	Full/Partial Shade		Evergreen, Groundcover, Good for shady slope bottom	Green-white	Low
Elymus canadensis	Canada wild rye	Yes	3–8	Full/Partial Sun	Yes	Meadow	Green	Tall
Koeleria glauca	Blue hair grass		5–9	Full/Partial Sun		Accent	Cream/White	Low
Koeleria macrantha	Prairie June grass	Yes	3–9	Full/Partial Sun		Accent	Cream/White	Low
Sporobolus cryptandrus	Sand dropseed	Yes	3–9	Full Sun	Yes	Meadow	Pink/Brown	Tall

* These species would benefit from another 2 inches (50 mm) of growing media but may survive in the shallower depth if regular irrigation is provided.
** Some of the *Sedum* species listed as evergreen may only be so in milder winters. Check with your local nursery.

Plants to Avoid

While tap-rooted and thirsty plants may not be suitable for a green roof, there are some plants that should be avoided altogether. These include noxious weeds and invasive species. Agricultural ministries or state departments create noxious weed lists because they affect crop production and can pose risks to the health and well-being of agricultural workers and livestock. Invasive species are usually defined as nonnative plants that potentially pose negative impacts on humans, animals, and ecosystems. Some of these may include:

- **Aggressive spreaders and seeders:** *Medicago* spp. (black medick); *Capsella* sp. (shepherd's purse); *Cerastium* sp. (chickweeds); *Digitaria* spp. (crabgrasses); *Alliaria* sp. (garlic mustards); and *Chenopodium album* (lamb's-quarters). While some of these may be drought tolerant, they will aggressively spread and outcompete all other plants.[2] Removing them is nearly impossible.
- **Plants with aggressive root systems:** Bamboos and *Polygonum cuspidatum* (Japanese knotweed) have very aggressive and strong root systems that could ultimately damage your underlying layers. Japanese knotweed is hard to eradicate and has taken over many natural environments.
- **Plants with harmful features:** This may include thorny or poisonous plants. While some plants on our list may have spines, i.e., *Opuntia humifusa* (eastern prickly pear cactus), these species are rare and native and encouraged on a protected green roof. Please make sure potentially harmful plants are somehow identified for anyone visiting or doing maintenance on the roof.

Always check with your local municipality or extension office to see which plants are restricted in your region.

Turf Roofs

Turf is an option for planting on roofs. Turf roofs are normally reserved for intensive green roofs with permanent access for visitors. Maintenance requirements are high (depending on the desired aesthetic), as they require regular mowing, irrigation, and nutrient amendments. Condo towers with recreational space, or parks and sports facilities over underground parking will often reserve a space for turf. However, some DIY builders have planted roofs with grass seed mixes with no intentions of mowing. See Photo 11 in the Color Section.

Moss Roofs

As discussed earlier, the environment on a green roof has elements similar to many natural environments. Moss has a common presence within the flora of many sites and often plays an important role. As a pioneering species, moss contributes to natural succession, making harsh environments more hospitable for other species. Therefore, moss may be able to contribute to biodiversity success on extensive green roofs.

The ability of moss to thrive in harsh environments is a result of some key characteristics. Most mosses can absorb dissolved minerals and substantial amounts of water, making these immediately available for photosynthesis. This characteristic allows mosses to get established on very poor or bare mineral surfaces and rock pavement. On a green roof, they can grow on very shallow substrates, with no supplemental irrigation or nutrition once they are well established. Many mosses have a high tolerance to drought; they will go dormant and brown out, turning green again upon receiving rain. Their structure allows them to capture sparse resources for themselves as well as neighboring species. Therefore, in environments where soil is limited, such as a green roof, both moss and herbaceous species are able to co-exist, thus leading to a natural evolution of species on the roof.

Often, moss is a natural colonizer on a roof rather than an intended species. This is partly because few are commercially available. A quick look through any nursery center will show you that moss is not widely available for purchase. Instead, you may need to seek out a moss-specific nursery such as Moss Acres in Pennsylvania. Otherwise, moss would have to be harvested from local environments. There is a negative stigma with harvesting wild plants. Doing so *can* remove vital components of an ecosystem that may take decades to replace. However, harvesting can be done sustainably if one does it sensibly, following the guidelines set out by conservation or ecological societies. I recommend looking around your own property for mosses, as you will likely find some.

Photo 12 in the Color Section shows a moss roof built by Moss Acres.

Moss is sometimes hard to establish due to its intolerance to metals. Ideally, moss would be watered with rainwater (harvested, if necessary) because copper and other metals found in city water sources are damaging to moss. Potable municipal water usually contains lime, which can create a suffocating build-up on mosses. If you are collecting rainwater, be aware that zinc-plated roofing material or copper roofs will leach these metals into the water collection system. Metal-tolerant mosses do exist, but you may be limited to what you can find.

Based on the literature and what I've discovered on green roofs, I've compiled the following list of moss I believe could work on a green roof. These species have been included because they have a wide distribution in both natural and man-made environments, so they should be able to adapt to the chemistry of most green roof substrates.

Moss Species with Potential for Use on Green Roofs:

- *Atrichum angustatum* (lesser smoothcap moss)
- *Bryum argenteum* (silvery sidewalk moss)
- *Bryum caespiticium* (sidewalk moss)
- *Bryum capillare* (capillary thread moss)
- *Ceratodon purpureus* (fire moss)
- *Dicranum scoparium* (broom moss, rock cap moss)
- *Encalypta procera*
- *Funaria hygrometrica*
- *Hedwigia ciliata*
- *Pohlia nutans*
- *Polytrichum juniperinum* (juniper haircap moss)
- *Polytrichum piliferum*

Fig. 6.4: *Moss began to form on this aggregate-based soil three years after it was planted. Ten years later, it's formed a crust that prevents weeds and wind erosion.* Photo Credit: Restoration Gardens, Inc.

- *Tortella tortuosa* (frizzled crisp-moss)
- *Tortula ruralis* (star moss)

When researching ways to establish moss, you will often see recommendations for mixing moss up into a milkshake compound with buttermilk or even with beer. The liquid mix acts as an adhesive to keep the moss from blowing away. You could also use netting or hydrogels. Photo 13 in the Color Section shows a test plot mixing moss with biochar and hydrogels.

Methods of Planting

Just like there are options for installing green roof layers, there are also options for installing plants. These include seeds, cuttings, plugs, sedum mats, or modular systems.

Seeding is a relatively simple and cost-effective way to grow a garden; however, it is not one that is widely used. The best time to lay wildflower seed in temperate climates is in the fall, when the temperatures are cooler, there is moisture available, weed seeds are dormant, and the sown seeds can settle down in the ground over winter. Seed mixes often contain a cover crop to cover the bare area in the fall, preventing erosion and reducing competition from weeds. In early spring, you can mow down the cover crop and leave it to provide nutrients to the soil; this is known as green mulching. See "Meadow Management" in Chapter 10 for more details.

Cuttings are almost exclusively used for planting sedum roofs. Nurseries that offer pre-vegetated *Sedum* products (mats and modules) usually trim the plants to keep growth low while the pre-grown material fills in. As a result, they have sedum cuttings available for their own use or to sell to others. *Sedum* roots down quite easily, so cuttings are a simple way to provide quick coverage if you don't have the budget for pre-grown mats or modules. Similar to seeding, you will at first need to ensure cuttings are kept moist, and, depending on your location and region, you may want to install protection from wind erosion.

Plugs are small plants with a pre-established root system in a small amount of soil. Plug containers are smaller than standard nursery pots, so they have smaller root systems. Depending on your location, budget, and timeline, you may be able to make good use of plugs, which will bring down your costs. Look online for nurseries that sell plugs specifically for green roofs. These

Fig. 6.5: *Installing Sedum cuttings can be quite quick, and they begin to fill in within a year. See Photo 14 in the Color Section to see this roof two years later.* Photo Credit: Restoration Gardens, Inc.

nurseries often grow the plants in media that is similar to what you will be installing on your roof. This is a benefit to your plants because they will not experience transplant shock as much as those plants that are grown in traditional fertile nursery soil.

However, many DIY green roofers will find themselves heading to their local nursery. If this is the case with you, here is what you should look for:

- Plugs or small pots. Large nursery containers are not recommended. Instead, look for 2–4 inch pots or plug trays. If you have deeper media, you could look for 1-gallon pots, but some of the root growth may have to be removed; such plants are accustomed to growing in fertile organic media, not green roof media, so they might not readily adapt.
- Clean stock with no weeds in the soil.
- Plants that require similar growing conditions to what they will have on your rooftop (i.e., similar sun exposure and growing zone).

Sedum mats are pre-grown rolls of *Sedum* plants rooted into a thin layer of growing media, sometimes with a piece of geotextile attached to prevent soil loss during install. They are very similar to the sod rolls you can buy at garden centers. And, just like sod rolls, pre-grown sedum rolls provide instant, mature coverage upon installation. In areas with high wind exposure, pre-grown mats stapled into the soil provide valuable coverage against wind erosion.

Modular units are available in a variety of options. Some manufacturers offer plastic trays pre-grown with drainage, geotextile, growing media, and plants, while others offer trays with growing media for you to plant, or even just the trays for you to fill. Typically, the trays come in squares or rectangles, 4–16 square feet per unit. Pre-grown trays are a good solution for sloped or windy roofs. In Figure 6.7, I compare some growing features of the various planting strategies. For modular units, I am referencing those that come pre-grown with plants. Photo 15 in the Color Section shows a roof on which we employed four of the five different planting methods for the clients: cuttings, plugs, pre-vegetated mats, and modular units.

Fig. 6.6: *Plugs are an ideal way to plant a green roof because they have small root systems. As some trays contain 72 plants, it takes no time or effort to carry nearly 100 square feet's worth of plants up to a roof. Involve kids on roofs that are safe with access and railings; they love the idea of putting plants on a roof!*
PHOTO CREDIT: LESLIE DOYLE

Figure 6.7: A Comparison of Install and Design Factors for Some Different Models of Planting Green Roofs

	Seeds	Cuttings	Plugs	*Sedum* Mats	Modular
Maturity for 80% Coverage	2–3 years	1–2 years	1–3 years depending on species	Immediate	Immediate
Amount Required	30/square foot	25–50 lb/1,000 square feet (12 kg/100 sq. meter)	1–2 per square foot (11-12 per square meter)	Complete square footage of vegetation roof space	Complete square footage of vegetation roof space
Required for Establishment	Carrier medium such as sand or cellulose to ensure even application; Wind erosion blankets depending on site; Irrigation if not planted in the fall	Irrigation	Irrigation	Irrigation	Irrigation
Species Diversity	*Sedum* Some wildflowers	All *sedum*	Full choice	Typically 20 pre-chosen *Sedum*. Initially higher, stronger competitors tend to dominate after a few years	Depends on system provider and depth. Stronger competitors tend to dominate after a few years
Design Limitations	Meadow or mixed *Sedum* only	Mixed *Sedum*	Unlimited design	Mixed *Sedum*, can add plug accents on roof	Mixed *Sedum* Mixed perennials, designs are costly
Cost	Low	Low	Moderate	High	Comparison is difficult as these contain all elements of the green roof
Labor	Easy	Easy	Easy	Easy, but requires heavy lifting	Easy, but may require heavy lifting
Delivery Considerations	Can ship in the mail	Fragile to ship; may need refrigeration	Long distance shipping requires refrigeration	Long distance shipping requires refrigeration. Have to call for availability. May require longer lead time	Long distance shipping requires refrigeration. Have to call for availability. May require longer lead time
Install Window	Fall is best	Requires time to root down before frost. Install immediately upon arrival	Requires time to root down before frost. Install upon arrival or protect and irrigate for a few days.	Longer seasonal window as roots are already embedded in growing media. Install upon arrival	Longer seasonal window as roots are already embedded in growing media. Install upon arrival or protect and irrigate for a few days if they are not stacked.

Figure 6.7: Continued

	Seeds	Cuttings	Plugs	*Sedum* Mats	Modular
Installation Tools	Seed Spreader	Rake	Trowel	2 or 3 people and potentially a lift	Minimum 2 people and a lift
Main Benefits	Cost Ease of application	Cost Ease of application Rapid coverage	Design Ease of application	Immediate coverage Flexible uses	Immediate coverage All-in-one system Removal ability

Plant Material Weight

When planning a green roof on paper, be sure to include the weight of the plants *at maturity*, not just at the plug or cutting stage. The following is a list of average plant weights at maturity:[3]

- Succulents, herbs, grasses: 2 lb/sq. ft (10 kg/m^2)
- Grasses and bushes up to 6": 3 lb/sq. ft (15 kg/m^2)
- Shrubs and bushes up to 3': 4 lb/sq. ft (20 kg/m^2)

Designing with Plants: Some Tips

There are many ways to design a garden, and many books have been written on just this topic. This is not one of those books, but here are some important tips for planting on a green roof.

When planning your design, consider the following:

- Plant spacing should be approximately 1 per square foot, more for sedums, less for grasses.
- If you cannot see your entire rooftop when standing at grade, considering planting some taller species—either throughout or just near the edge. You might even wish to include some plants that will eventually cascade over the edge.
- If you can see your green roof from a window, put some consideration into where the taller species sit so as not to block your view of the others.
- Consider a diverse planting plan. This will make your roof less susceptible to drought and disease and increase the biodiversity value of your green roof (see next section). It will also extend your roof's visual interest throughout the growing season.
- If planting on a slope, place plants with higher drought tolerance at the top.

Biodiversity on the Green Roof

Biodiversity is integral to species survival. A diverse planting will limit the stress of drought and disease on a rooftop far better than a monoculture of plants. The City of Toronto produced a guideline for encouraging biodiversity in which they state that vegetation and site diversity increases the opportunities for pollination and food, shade, nesting, perching, and nutrients to visiting birds and insects.[4] In addition, ecosystem services increase with biodiversity.[5] Therefore, any way to increase diversity is of value. Even choosing multiple species of *Sedum* extends the flowering period for pollinators.

Ways to increase biodiversity are not complicated, as seen in Figure 6.8, and the various methods can be executed on both extensive and semi-intensive green roofs. Some of these methods were employed on the roof shown in the Color Section in Photos 16a–c. Strategies include:

- Incorporation of variable depths of growing media, i.e., adding some height around point loads or edges.
  - *Benefit:* Allows for a greater diversity of plant species.
- Creation of small soil mounds and divots.
  - *Benefit:* Creates microclimates with areas that are dry and exposed, as well as those that are sheltered and moist, increasing the suitability for more diversity.
- Incorporation of variability in the growing media composition, i.e., some areas with smaller particles and some areas with larger particles.
  - *Benefit:* Creates habitats for insects and soil nematodes.
- Incorporation of logs, rocks, or branches.
  - *Benefit:* Allows for perching and nesting; creates small microclimates.
- Incorporation of pitted rocks or shallow dishes with pebbles.
  - *Benefit:* Creates small water sources for visiting birds and pollinators.

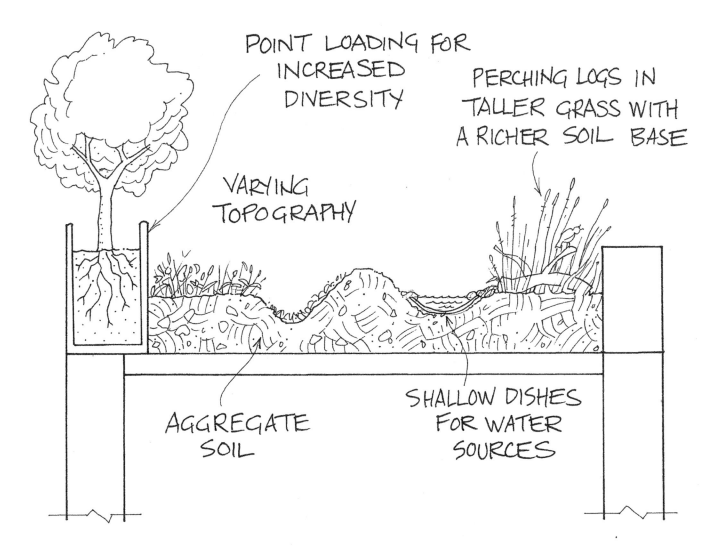

Fig. 6.8: *Some strategies can be engineered, such as taking advantage of point loads, while others are as simple as creating divots and variation in growing media.*

Designing with Maintenance in Mind

Some green roofs will require more maintenance than others; but, as any gardener knows, the amount will largely depend on your aesthetics. A weed is only a weed if it's undesirable in your own garden (or your neighbor's). Weeds can be introduced onto your roof or taken off your roof by wind, birds, or even the bottom of your shoes. And, while you may not mind the additional flowers of some blow-ins, some species can choke out your intended species, compete for the little nutrients and water that are available in the growing media, then die out in droughts, leaving unappealing brown gaps in the roof garden.

Here is a simple scale of maintenance for green roofs:

- **Highest:** Vegetable production or manicured landscape. With these green roofs, maintenance activity will be frequent to remove weeds and ensure that the desired plants are in their intended pattern, with optimal growth. If self-seeding plants are included in the design but spreading isn't desired, flowers will have to be cut back before they set seed or seedlings will have to be removed. Irrigation will be required on these roofs, and, as a result, the growing area will be hospitable to many unintended blow-ins.
- **Moderate:** Native or mixed plantings. Care is required 2–3 times a year for the pruning of seasonal growth and removal of weeds. Irrigation may only be required in times of drought, and therefore temporary hand watering may be possible. Natural competition and die-out are tolerated, to an extent.
- **Low:** *Sedum* or naturalized gardens. *Sedum* roofs typically achieve full coverage in 2–3 years, leaving very little space for weed blow-ins. They require little nutrition or supplemental water once they are established. To encourage vigorous growth, remove brown spent seed heads in spring. As mentioned earlier, clippings can be left in place to root down or replanted to fill in gaps. This is best done in the early spring before new growth occurs and to take advantage of spring rains for the cuttings to root. Naturalized gardens develop when roof owners accept natural competition, blow-ins, and die-outs. Some species (intended or not) may outcompete others, resulting in the disappearance of other, less competitive species after the first few years. Irrigation is not required after establishment; therefore, the vegetation may experience periods of dormancy or brownouts, with the regeneration of opportunistic species. In this case, the goal is to ensure a thriving community of plants, regardless of species. Please note that owners who intend on cultivating a naturalized green roof still need to visit the roof once a year to remove any tree seedlings or restricted species and to check on roofing components.

When to Plant

There are two times that are ideal for planting a green roof:

1. After last frost, to take advantage of spring rain and cooler temperatures.
2. In early fall, to ensure plants are rooted thoroughly before the first frost.

Search online for your first and last frost dates, if applicable, to determine your planting window. Plants should be put in about a month before first frost to give them time to successfully root down. In mild and moist winters, like those on the Pacific Northwest coastline, you have the option to plant right into the winter. If you must plant in the summer, ensure the roof receives daily irrigation.

Climate change is resulting in greater anomalies in the weather. Our seasonal temperatures are not always reliable year to year, so it is best to be conservative and not push your planting window.

Establishment Period

Regardless of your installation method, adequate care is required after planting to ensure that plants survive in their new environment. This is the plants' *establishment period*. The industry's definition of establishment period is the time it takes for vegetation to reach 60–80% coverage. Your first goal, however, will be to ensure your plants survive the transplanting. During this time, plants will need to root into the growing media; in order to do this, they need lots of water. Therefore, the time it takes to establish the plants could be 3–4 weeks if temperatures are mild and adequate water is provided, or longer if the plants are exposed to high, stressful heat, which increases evaporation. *Sedum* roofs, regardless of when you plant, usually root down in 3 weeks. It is best to water your garden daily during this period. After this daily regimen, you can start reducing the amount of water each week, starting with alternating days and then reducing further from there. This will wean the garden off water.

For the first summer, ensure that the roof is watered during heat waves, as the plants will not have established their drought tolerance yet. While it may be tempting to pamper your roof with water and nutrients year after year, this is not sustainable, and it may do more harm than good. Plants that exhibit lush growth may not be as drought tolerant as plants that still continue to grow but show stunted growth.

Now that you have established the type of garden you can build on your roof and the plants you wish to grow, we will explore the various materials you can use to construct your system.

Chapter 7

Green Roof Material Options

A ROOF IS EXPOSED to many damaging elements. UV radiation from the sun can break down a membrane. In the winter, snow and ice can bend flashing, pulling apart seams or joints, and melted water entering a tiny crevice can quickly freeze and expand, causing further stress on vulnerable areas. On a daily basis, the temperature fluctuations that occur from day to night can cause materials to shrink and contract, placing constant stress on connections and seams. By covering your membrane with a green roof, you are reducing its exposure to these elements and thus expanding its lifespan. In many cases, green roofs can double the lifespan of a membrane.

How "Green" are Green Roofs?

While green roofs are seen as a low-impact design strategy, many products in the layer systems do have large carbon footprints. Most membranes used in a green roof would have been used in the same flat roof had it not been greened. However, extra elements such as stronger structural members, poly sheeting, and aggregates do require energy-intensive harvesting, manufacturing, and shipping resources, which invariably contributes to habitat and land degradation. These emissions are often counterbalanced by a roofing system that reduces run-off, increases thermal performances of buildings, reduces the urban heat island effect, filters out particulate matter, or even produces food.

Studies have attempted to determine the point at which a green roof becomes a carbon sink—after accounting for the amount of emissions that came from the products used to make it. While this point greatly depends on the materials and quality of the build, it could be anywhere from 15 years[1] to two-thirds of the roof lifespan. The best you can do is install a quality membrane and preserve its integrity so that none of your efforts or materials go to waste. Look for materials with high recycled content. While processing recycled materials still emits carbon and other harmful gases, it has a lower footprint than the extraction and processing of virgin materials.[2]

In this chapter, where possible, I will give embodied carbon amounts of common green roofing materials, as well as any other significant environmental notes as they pertain to the different material choices. Note that all data on embodied carbon is from the *Inventory of Carbon and Energy (ICE) 3.0*, University of Bath, Nov 10, 2019.

Required Layers

As previously mentioned, there are essential and optional layers. Essential layers include the waterproof membrane, root barrier, drainage layer, filter cloth, growing media, and plants. Optional layers are based on the needs of your design and include membrane protection, insulation, water retention, slope restraint, wind erosion protection, and irrigation. In this chapter, we will highlight the role of each layer and the types of materials available. For the most common layers, I will include saturated weight loads, costs, and embodied carbon.

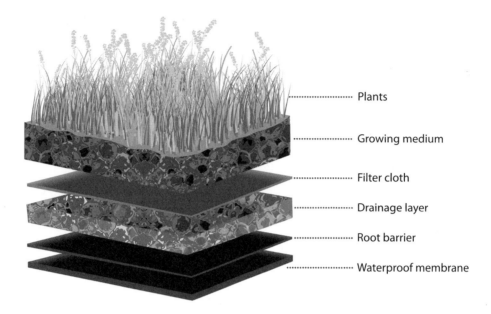

Fig. 7.1: *Required layers of a green roof.*

Fig. 7.2: *Roofs with leaks should be thoroughly investigated rather than just recovered; you want to determine the extent of any water damage.*
PHOTO CREDIT: RESTORATION GARDENS, INC.

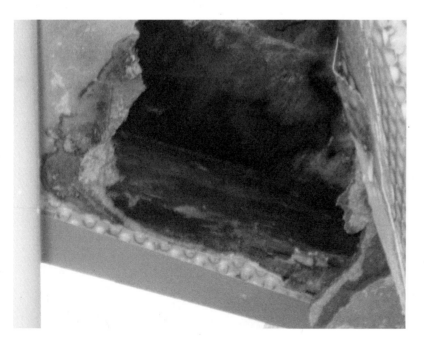

Waterproof Membrane

Role

The only job for the waterproof membrane is to keep your roof watertight, free from any leaks. Depending on the scale and complexity of your project, this may be the one layer that you hire a competent expert to install since once a green roof is installed, finding the source of a leak can be an expensive venture.

If you are retrofitting an existing roof, consider hiring a roofing company to examine your roof for any areas of concern, especially if it is over five years old. A roofing expert could perform a leak test or do a thorough examination for vulnerable points. Sometimes if there are no leaks, but the existing roof is showing small signs of wear, you can recover the existing roof with a compatible one.

If your roof has experienced multiple leaks in the past, it is best to replace the membrane. Recovering in this situation may seem like the quick answer, saving you the hassle of tearing off the old membrane and sending it to the landfill. However, while you may have solved the leak, you will not have investigated the extent of any structural damage. The structural deck could be jeopardized or perhaps the underlying insulation is saturated, which would negate any thermal properties.

Waterproofing Assemblies

A waterproofing assembly, which creates a watertight seal, is recommended over a roofing assembly, which only sheds water. Shingles are an example of a roofing assembly.

Waterproofing assemblies over insulated roofs include conventional roof assemblies and protected membrane roof assemblies (PMRA). The former locates the insulation under the membrane, while the insulation in the latter assembly is placed above the membrane. Both are suitable for green roof applications. Roofs over uninsulated spaces, such as porches or sheds, can be suitable for a green roof; however, the plant choice needs special consideration.

Membrane Installation Methods

There are two ways to install a waterproof membrane: fully adhered or loose laid. Fully adhered membranes use adhesives or solvents to adhere the membrane onto the deck. These membranes are hot-applied, cold-applied, torch-applied, or self-adhering. Loose-laid membranes are draped over the deck and held down via fastening systems or the weight of the green roof layers.

The benefits of fully adhered membranes include:

- Elimination of lateral movement of water between the membrane and the roof deck, should a leak occur. This makes it easier to locate a leak.
- The membrane will have greater resistance to wind uplift.

The benefits of loose-laid membranes include:

- Quick and cost effective install.
- Material can be lifted and recycled or separated for disposal at the end of its life, keeping the deck intact.

For loose-laid applications, the green roof layers must be installed immediately to weigh down the membrane; otherwise,

Leak Detection Tests

Leak detection tests can provide you confidence that your membrane is watertight. There are different ways in which a leak detection test can be executed, but the easiest one—if applicable to your roof design—is the flood test. One version of a flood test can be carried out with parapet walls, the other without.

This first flood test should only be performed on flat roof with slopes up to 2%. Perform the test with at least two people; one person to inspect the underside of the roof and one person on the roof managing the flow of water and watching for signs of leaks.

Steps for a simple flood test:

1. Plug the drains using temporary patches or sandbags and plastic. In one project, we used a product called Black Knight waterproof bitumen repair sealant that can be applied in wet conditions and is slow drying. This allowed us to perform the test and then remove it upon completion. Make sure your patch is supported from behind (with lumber or rigid foam) so the force of the water cannot push it open.

2. Flood the roof with 2 inches of water (or more—on a slope; you need enough to have 2 inches at the highest point), again ensuring your roof can handle the weight. Be conscious of the fact that for each inch of water, you are adding roughly 5 lb/ft^2 (2.4kg/m^2) of weight. As the roof is filling, check for any air bubbles or directional water flow and have someone under the structure in a safe location inspecting for

any infiltration (this will not be immediately seen if you have a finished ceiling with insulation). Check the outside of the scupper for water seepage from a poor patch. Do not leave the roof unattended until you are confident there are no immediate threats. If you notice any indications of leaks, immediately remove the patches and encourage the roof to drain so the large volume of water does not overwhelm the leak source.

3. Mark the height of the water (on a ruler or on the inside of the parapet wall).
4. After 24 hours, check the height of the water. If it has not lowered, you have a watertight roof. If the water level has changed, you need to find the source of the leak. Always check around your sealed drain outlets first, as your seal may have failed.
5. Slowly remove the patches so that water begins to exit the roof without overwhelming the drainage system.

Flood tests should not be executed on roofs with higher slopes, as achieving an inch or two over the entire roof can result in the lower end of the slope having to support too much weight. In these cases, you could execute a *flowing-water test*.

In the flowing-water test, water flows continuously over the surface for a minimum of 24 hours. No drains are blocked in this application, so this may be a suitable test for roofs without parapets. Note that it is challenging for a flowing water test to pin point the source of water entry.

Flood testing is not always feasible, but there are alternative tests available. These require hiring a professional, which will add to the cost of your project—but they can ease your mind about leaks.

Other tests, which can be performed by professionals, include:

- Electronic leak detection (low voltage or high voltage)
- Impedance tests
- Infrared (IR) thermal imaging

Fig. 7.3: *A roof covered in water for a flood test.*
Photo Credit: Restoration Gardens, Inc.

temporary ballast will be required. Note that in loose-laid applications, in the case of a leak, the lateral migration of water makes it difficult to spot the source.

Refer to your local code to determine your allowable installation method, as some regions may stipulate that you must fully adhere the membrane under the green roof.

Be sure to use compatible flashing material (more on this, later in this chapter), and after the work is complete, it is recommended that you perform a leak detection test. Check with your local municipality to determine if one is required.

Specifications

There are four recommended types of membranes, but there are no specific requirements to use any of these[3,4]:

- Polymer-modified asphalt membrane, nominal 215 mil (0.215 inch, 5.375 mm) minimum thickness, fabric reinforced, hot-fluid-applied.
- APP or SBS—polymer-modified bitumen sheet membrane, two-layer minimum or 145 mil (3.7 mm).
- Ethylene propylene diene monomer (EPDM) membrane, 60 mil (0.06 inch, 1.5 mm) minimum thickness, reinforced, with stripped-in lap seams.
- Fluid applied elastomeric membrane, fabric reinforced, one- or two-component.

It would be advisable to look for membranes that have a good track record in your local climate. Membranes must have resistance to changes caused by the chemical and biological action of microorganisms as well as chemical components existing in the vegetation and growing media or dissolved in the water. (So, this includes the chemicals you might introduce via organic and synthetic fertilizers.) The membrane should also be able to adjust to the seasonal movements of the roof during the freeze/thaw cycles, and even though the green roof layers cover it, the membrane should have good UV resistance. Ask manufacturers for the lifespan of the membrane, both covered and uncovered by a green roof.

Below, I highlight three of the products a DIY builder is most likely to encounter: modified bitumen, single-ply synthetic rubber, and liquid-applied products.

Fig. 7.4: *Installing a 2-ply mod bit membrane.* Photo Credit: Restoration Gardens, Inc.

Modified Bitumen

Modified bitumen membranes (aka mod bit) are asphalt-based and very common, and there are lots of options with this material. The DIY builder should look for styrene-butadiene-styrene (SBS) modified bitumen. These are more flexible and have options that can be adhered or self-adhered.

Self-adhering systems are good for lower slopes between 1:12 and 4:12. If you have a roof with parapets, install cant strips to reduce the internal angles. Sheet sizes are typically 39.375 inch (1m) and consist of 1–3 plies, including a base and a cap sheet.

Environmental Notes

- Bitumen is a crude oil product, and these membranes also contain polymer modifiers, filler, and additives that are reinforced with glass fiber and/or polyester.
- Some have higher recycled content, which reduces virgin material resources.
- Self-adhering systems give off no fumes or odors during application, unlike hot asphalt or solvent-based applications.

Single-Ply Synthetic Rubber

Single-ply membranes consist of large flexible sheets that are rolled out onto the roofs. There are many versions available, and all are petroleum-based. Some single-ply options for green roofs are PVC, TPO, and EPDM. PVC and TPO are root resistant; so is EPDM if there are no seams.

There are a variety of products available with varying widths and application options. There are even some self-adhering TPO membranes available. Seam work can be difficult to execute. This is where having some experienced help is particularly useful.

There are mechanical fastening options for all; otherwise, PVC and TPO require heat welding or solvents for seams, while EPDM requires only adhesives. Always review manufacturer literature for best practices. Due to the many available roll sizes of EPDM, it is often used in DIY installations; it can easily be draped over the entire roof and held in place with the green roof ballast.

Rolls of single-ply membranes typically come in 7.5, 10, 16.7, 20, 30, 40, or 50-foot widths and are 100 feet long. You will have to buy whole rolls unless you can buy leftovers. You could instead try finding a garden center that cuts rubber pond liner to length. But note that most EPDM pond liners are thinner, at 45 mil instead of 60 mil, so you may want to use these only for an extensive system or double them up for a semi-intensive or food system to increase its durability. Always inspect membranes for damage. If you are loose-laying the EPDM over your roof, it is recommended that you install it over a fleece layer (or, you can use extra geotextile) so that it does not snag on any sharp edges. Buy a little more than is required so you can use cutoffs to cover any exposed sections, creating a double layer of protection against UV damage. Areas that need protection are typically at the edge of the roof. See Figure 8.9, which shows a sacrificial layer.

Liquid-Applied Products

Liquid-applied membranes, whether sprayed or painted, hot or cold, offer a seamless application applied directly to the structural deck. These can be used on low-slope or high-slope roofs and on roofs with or without parapets.

There are many products available on the market. Talk to the retailer or manufacturer about using it in a green roof application for best practices and product lifespan. If you plan on spraying overtop an existing membrane, check the manufacturer for compatibility.

Fig. 7.5a and b: *EPDM patches and seals should be executed as per manufacturer's instructions and with proper EPDM adhesives. The top photo shows a tight EPDM corner seal, while the photo on the bottom shows an EPDM patch that has failed.*

Environmental Notes

Look around for water-based membranes, which contain no solvents or VOCs.

In Figure 7.6, some application information for the three recommended materials is compared.

Figure 7.6: Application Comparisons for Some Green Roof Membranes

Advantages	Mod Bit	EPDM	Fluid-Applied
Applied on Conventional and PMRA	X	X	Ideally PMRA
Root Barrier	Required	Not required unless there are seams	Required
Recommended Application for DIY	Self-adhered or mechanically fastened	Loose lay/ballast	Paint or spray
Ease of Install	Easy to hoist, corners and details can be tricky. Can find easy to follow instructions including fastener spacing and how to handle flashing	Easy, just roll it out. It's heavy so getting on the roof is hard. Corners and details can be tricky.	Easy, especially for handling penetrations and corners, but need to ensure even application
Seams	Easy to manage	Harder to manage, look for a larger piece of material to avoid seams	No seams
Longevity (confirm with manufacturers)	Moderate	Highest	Lowest
Slopes	Typically used on low slopes	Do not loose lay above 17%; Do not use above 57%	No limits
Install Window	10–35°C	Anytime unless using seam adhesives, then above 7°C	Above 5°C Requires time to cure, sometimes 48 hrs.
Water Migration	Low if self-adhered	High for loose laid	Low
Availability	Most hardware stores	Roofing or building suppliers	Most hardware stores
Material Properties	High puncture resistance, seam strength and UV resistance	High UV resistance; good puncture and seam strength; incompatible with petroleum products (cannot lay overtop of bitumen membranes or contact with petroleum seals, or oils and grease such as roofs with restaurant vents) Not compatible for EVFM leak detection	High freeze/thaw resistance, UV resistance, adhesion and chemical resistance
Embodied Carbon*	6.7 kgCO$_2$e per m^2 of membrane	EPDM: 6.69 kgCO$_2$e/m^2 TPO: 4.42 kgCO$_2$e/m^2 PVC: 5.3 kgCO$_2$e/m^2	9.79 kgCO$_2$e/m^2
Saturated Weight Load	A 2-ply system is approximately 1.24 lbs/sq. ft	60 mil EPDM is 0.43 lb/sq. ft	0.75 to 1.5lbs/sq. ft [5]
Cost	A 2-ply system can start around $2/sq. ft for the sheets only, not including any necessary fasteners or tools	Approximately $1/sq. ft for the membrane only	Approximately $3.50/sq. ft; Look at the manufacturer's websites for coverage rates

* Data is on embodied carbon is from the *Inventory of Carbon and Energy (ICE) 3.0*, University of Bath, 10 Nov 2019.

Amount Required

Membranes should cover the total area of the green roof as well as the vertical surfaces up the face of the parapets and connecting walls. They are usually terminated on the top of the parapet and at the height deemed appropriate by code on the walls. If you don't have parapets, the membrane extends to the edge of the deck or slightly over it.

Root Barrier

Role

The role of a root barrier is to protect the integrity of the membrane from root growth. Most plants that are suggested for a green roof have fibrous root systems that spread out horizontally versus a single taproot; however, even small roots can cause damage if allowed to grow into a membrane. If the membrane itself is not root resistant, a separate layer is required. In this case, the layer should be installed over the *entire* surface of the membrane, including the vertical surfaces.

Specifications

Root barriers tend to consist of solid, flexible sheets of plastic. They should be UV-resistant; if not, they need full protection against sun over their entire surface.

For semi-intensive asphalt or bitumen roofs, a minimum of 0.03 inch (0.8 mm/30 mil) should be used. There should be no seams, or seams should be lapped by 6 inches, with a welded seam a minimum of 1.5 inches wide.[6] A welded seam is created by using heat to seal the sheets closed.

For extensive systems, a thinner profile 0.50 mm/20 mil is sufficient; seams must be welded.

Heat welding is hard to execute. A DIY builder might be better off looking for large sheet material to cover the entire surface or using a flat-profile barrier with at least 5 feet of overlap. ZinCo is a green roofing supplier that offers root barriers as well as welding kits.

Products

Root barrier products include:

- High-density polyethylene (HDPE)
- Polypropylene (PP)
- Copper sulfate impregnated fabrics
- Copper foil
- EPDM, PVC, and TPO roofing membranes

When choosing your components, consider their compatibility with one another. If you are using a root barrier with copper-based compounds, those chemicals may leach into water or come in contact with

Fig. 7.7: *Root barrier should cover the entire surface.* Photo Credit: Restoration Gardens, Inc.

your membrane. Check with local environmental laws to see if this option is allowed. Make sure your root barrier can safely come into contact with fertilizers, humic acids (from growing media), and bitumen.

Root barrier is not the same as weed cloth from the garden center. Weed cloth is intended to prevent weeds from growing *up*. It contains perforations for water flow that roots can make their way through. Weed cloth is better used as a geotextile/landscape fabric layer above the drainage layer.

Embodied Carbon

HDPE: 0.98 kgCO$_2$e/m^2 at 30 mil

Polypropylene (virgin): 4.98 kgCO$_2$e/kg

For roofing membranes, see the table in Figure 7.6.

Saturated Weight Load

Consult the manufacturer's data. If it is a lightweight plastic that does not hold any water in cups, then assume it will not add significant weight.

Cost

Approximately $0.30 to $0.50/sq. ft for HDPE or polypropylene sheets, with various roll sizes available.

Amount Required

The amount of root barrier will match the membrane, as you are to cover the membrane completely, plus whatever you need for overlaps.

Drainage

Role

There are two components being referred to when we talk of drainage: 1. the designed elements that allow water to drain off the roof, and 2. a layer located between the root barrier and the filter cloth that allows water to move through the system.

They work together to prevent the over-saturation of the growing media; overly wet media can cause root rot as well as excessive weight on the roof. Note that you may not even require a drainage layer on roofs that have a slope higher than 5%. The angle of the

Fig. 7.8a and b: *On the left, water flows off the green roof via linear overflow, draining through a continuous opening at the edge. On the right, slopes in the roof deck direct water toward an internal drain kept free from clogging by the use of a non-vegetated zone.*

roof will move water freely toward the drains and off the roof.

Roof drainage elements include:

- Roof scuppers or spouts, including emergency overflow scuppers
- Linear overflow
- Interior drains
- Gullies or channels in front of doors
- Gutters

Specifications

The drainage layer needs to be:

- porous—to allow water to flow through and toward the drains.
- permanent—it must not lose drainage capacity or clog.
- continuous—it must cover the entire surface of the green roof.
- strong enough—to resist compression from the overlying layers.

Roof drain elements should be made of corrosion-resistant materials, and you should use corrosion-resistant accessories. Scuppers and drains need to be sized to accept the rainfall as required by codes, and they must be accessible for maintenance. Efforts need to be made to ensure that drains and scuppers do not clog.

In semi-intensive or food roof systems, place some drain boxes (sometimes called "inspection chambers") with removable lids around drains and scuppers. These keep the

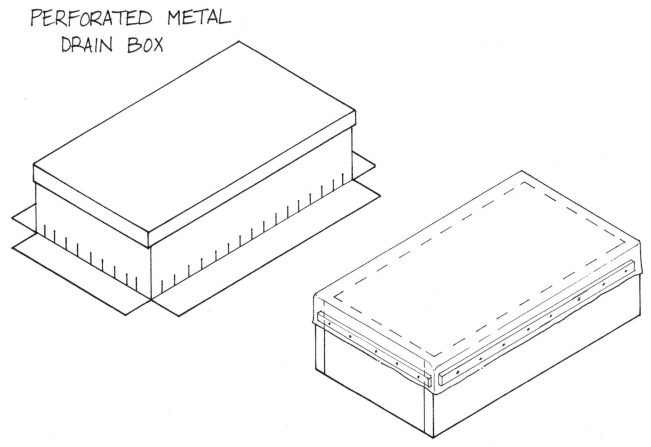

Fig. 7.9: *Drain boxes can be made of perforated metal with lids, or you can make one with composite lumber and cover it with filter cloth. If the latter, make sure it sits on top of the drainage board so water can flow underneath.*

Fig. 7.10 a,b,c:
*Drainage layers may be made of many different materials.
The photos here show
(a) aggregate;
(b) drainage cups; and
(c) dimpled board.*
Photo credit a and b: Leslie Doyle;
Photo credit c: Jayne Miles

areas free of debris to ensure free-flowing water can escape. The chambers and lids can be made of plastic or weather-resistant metal. They should have perforations on the sides for water flow, or they should be set on top of the drainage layer. Otherwise, a non-vegetative zone with clear stone, separated by edging or filter cloth, can be used around these areas so that water can flow off the roof freely, past the growing media. Clear stone is washed and contains no sand or fillers.

Emergency overflows are outlets in the roof design that are higher than the roof deck and main scuppers but low enough that once the system hits saturation capacity, water is able to flow off the roof. FM Global, a large risk management/insurance company, suggests designing the emergency overflow to be 2 inches (but not more than 6 inches) above the base of the roof deck. These overflows are important in systems with parapet walls to ensure that water can drain from the roof if the main outlet is blocked. This situation should be avoided at all costs, but it can happen if debris piles up after a storm or autumn leaves collect in the drains.

You must also prepare your green roof for any water that drains onto it from overhanging roofs. This contributed water should be distributed rather than left to pool around plants, as it could lead to root rot. A non-vegetated area or patio stone can be placed under downspouts to direct water into the green roof.

Products

Drainage layers consist of either 1–2 inches of coarse, clear (no fines), rounded aggregate or synthetic sheets, or a combination of these elements. Many are recycled HDPE, aggregate, nylon polymers.

If using aggregate with a single-ply membrane, consider installing a membrane protection sheet.

Synthetic sheets provide drainage in different ways, based on their design. One type, made of tangled polymer fibers or heat-welded foam pieces, allows water to freely move through it; other types retain some water in cups or depressions, before the water flows through the system. Often, sheet materials come with geotextile attached, which makes installation easier and prevents the geotextile from shifting.

Dimple board is a product used in construction as foundation wrap to prevent water from seeping into interior spaces. It is dimpled to allow the wall to breathe and

7.10 a: aggregate

7.10 b: drainage cups

to channel the water. This type of sheet material can be used as a drainage board on a green roof. You can find a dimple board with attached filter cloth at building centers that sell foundation wrap or sheet drain for construction projects. While the dimples in some types are in an orientation that does not allow for retention of water, you could turn it around and place the cloth on the other side, or just situate it as it comes, and it won't collect water. The table in Figure 7.11 gives some advantages of the various types of drainage layer options.

Roof drain components can be found online, at any roofing supply store, or at large hardware stores. Drain boxes specifically made for green roofs can be found online, or you can have your own made.

7.10 c: dimpled board

Figure 7.11: A Comparison of Properties between Options of Drainage Layers

	Aggregate	Sheet System with No Retention	Sheet System with Retention
Water-Holding Capacity	Aggregate will naturally retain some water	None, unless there is a thick fleece attached	Some retention—depends on the size and orientation of the depressions
Advantage for Plant Roots	Roots are able to spread out and grow in the voids	Beneficial, if roots are allowed to spread throughout the voids of the sheet	Provides some water for roots to take up after rainfall
Variety/ Availability	Lots of varieties available based on location. Some forms such as lava rocks or crushed bricks have high water-holding capacity and low weight loads	Dimple board products available at building centers	Limited based on region, may have to order from suppliers
Ease of Install	Labor intensive; must ensure even coverage	Quick roll-out install	Quick roll-out install
Saturated Weight	Can exceed 4 lb./ft² for every inch of depth	Lightweight; contact manufacturer for weight	Depends on the depression size, as well as the slope; contact manufacturer for weight
Cost	$50–100/cubic yard for pea gravel, which would cover approx. 350 sq. ft at 1"– approximately $0.25/sq. ft	$0.25~$1/sq. ft	~$0.25-$1/sq. ft
Embodied Carbon*	0.00438 kgCO₂e/kg for common aggregate such as gravel, pea gravel	Virgin polypropylene: 4.98 kgCO₂e/kg HDPE: 0.98 kgCO₂e/m² at 30 mil	For a layer of fleece 100% virgin polypropylene: 3.96 kgCO₂e/kg

* Data is on embodied carbon is from the Inventory of Carbon and Energy (ICE) 3.0, University of Bath, 10 Nov 2019.

Embodied Carbon

See Figure 7.11 for embodied carbon of different drainage layers. Some components might be made of recycled aluminum, which has embodied carbon of 5.65 kgCO₂e/kg (for 31% recycled content in North America).

Amount Required

The drainage layer should cover the horizontal surface of your entire green roof.

Filter Cloth

Role

A fabric made of polypropylene or polyester protects the drainage layer from filling with growing media or debris. This filter cloth is vital for the integrity of the drainage system. It must be permeable and penetrable by roots, allowing them to access air space in the drainage layer or additional water in the water-retention layer.

Specifications

Both nonwoven polypropylene and polyester fabric should be of a weight between 3 and 6 oz/yd³ (100–200 g/m²).[7] Try to find fabric with good UV resistance as this layer may be exposed in some areas.

A nonwoven geotextile looks like felt, whereas a woven geotextile looks like plastic with a weave pattern. Nonwoven geotextiles should be used because they have higher water permeability rates and do not lose strength over time unless they break down under UV.

Products

Filter clothes can be a separate layer or integrated with a manufactured drainage board; if joined to drainage board, the filter cloth is typically heat welded onto it.

Combination drainage boards/filter cloths are easier to install (though they can be hard to find), and they are sometimes more cost-effective. Typically, we cover up the vertical surfaces of the parapets with filter cloth. Therefore, if you decide on a combination sheet, you will need extra filter cloth for covering up drainage openings and/or installing around edges and termination points.

Most construction supply stores can offer you a custom cut of filter cloth from their large rolls. If you want to use thinner landscape fabric from a nursery, you can double it up, but it is best to get rolls that are as large as possible, so you reduce the number of seams and overlaps—these quickly become trip hazards, and they can be caught by the wind, creating a parachute effect.

Some thermoplastic (e.g., polyethylene or polypropylene) cloths are made from recycled material. Any recycled product will have lower embodied carbon than newly produced materials.

Embodied Carbon

100% virgin polypropylene: 3.96 kgCO$_2$e/kg

Saturated Weight Load

Varies; contact the manufacturer.

Cost

$0.10–$0.25 per square foot.

Amount Required

Filter cloth should cover the drainage layer. Be sure to cover drainage openings, soil restraints, and edges. I typically extend the layer up the vertical parapet as well to cover the exposed root barrier. By doing so, I am able to hold the filter cloth in place by the flashing.

Growing Media

Role

Growing media supports plant life by providing an anchoring layer in which plant roots can spread to absorb the water and nutrients contained within it. The growing media absorbs water for plants to take up in between rain events but also allows water to drain, reducing the weight on the roof structure. Most green roof growing media has both organic and inorganic matter, mostly in the form of aggregates.

The science of developing an industry spec'd engineered growing media is complex. Many physical properties are tested in labs to achieve standards for consistent and successful mixes. This book is not going to teach you how to reproduce an engineered growing media; instead, it will guide you toward choosing materials available in your own region for a custom mix. If you are within distance of a industrial supplier of pre-mixed growing media—and if it's within your budget—I would suggest using it; engineered growing media has much higher success rates than home mixes.

Specifications

Growing media components should be chosen based on four factors: climate, function, regional accessibility, and sustainability.

North America has vast differences when it comes to climate, so growing media will vary by region. In areas with frequent heavy rainfall, materials should have faster drainage capabilities. In areas that are hot, dry, or more susceptible to droughts, material should retain more water and contain components that can keep the root zone cooler. As moist conditions will accelerate organic decomposition, a growing media with a slightly higher organic content may be required in areas with high humidity versus areas with arid climates.

Growing media components should also be chosen based on the desired performance function. For example, a roof with a greater slope will require media that has more water-retention capabilities near the ridge and is more free-draining near the edge. A roof that is designed for stormwater retention will require porous water-retaining material, potentially with a lighter saturated weight. A roof in an area with high winds will require material of larger particle sizes with a higher dry weight to resist wind erosion. If you want to support a diverse plant palette, your media will need to offer more nutrition and water than a basic *Sedum* roof requires—more organic matter is the usual solution. A *Sedum* roof, on the other hand, fares better with free-draining, aggregate-based media.

Choosing materials that are available locally will keep costs down and is a more sustainable approach. Shipping large amounts of aggregate material can use a lot of fuel. Try to look for materials that are byproducts, attained through sustainable harvesting methods, or left over from other projects. Do not use products that leach toxic chemicals.

To make an ideal blend, ensure that your mix can do the following:

- Provide aeration for roots and resist compaction.
  - Provide a mix of particle sizes—⅛ to ⅜ inch with minimal (<25%) fines. If you must use sand, look for concrete sand (coarse) instead of play sand (fine).
- Provide a source of nutrients to the plants
  - Include a portion of organic material.
- Balance drainage with water-holding capacity.
  - Include porous aggregates that can hold water while also creating voids that allow for drainage.
    - Limit fines to reduce clogging.
- Resist decomposition.
  - Have a higher ratio of inorganic to organic material.
- Resist acidification via a high buffering capacity.
  - The ideal pH range for most plants is 6.0–8.0. If the growing mix contains more clay than sand, it will be able to better buffer against the acidification of acid rain. Limestone chips can also help buffer pH.
- Resist high winds as well as floating and displacement in heavy rain.
  - For areas susceptible to floods and high winds, include a higher volume percentage of material with higher dry weights and/or particle size; avoid material such as vermiculite and perlite, that can float in heavy rains.
- Reduce risk of fire.
  - Potting soils that include peat, shredded wood, bark, Styrofoam, or vermiculite are flammable. These should be mixed with nonflammable aggregates, so that mix is fireproof when dry.

Products

Most commercial green roof media mixes are made in bulk by specialized soil mixing companies; they are not found at your local garden store in bags. A quick internet search will point you in the direction of a distributor. Ask if they will deliver it to you in cubic yard bags or in bulk. Some soil mixers may require that you to pick up the order if only one yard is requested. If you can't get a specialized mix from one of these companies, you will have to mix your own.

At the beginning of the green roof movement, green roof media was typically 80% inorganic aggregate with 20% organics (by volume). "Larger" lightweight aggregates made up the bulk of the inorganics ratio, with coarse sand or fine, lightweight aggregate making up the "fines." The aggregate portion is what satisfied most of the requirements for growing media. A high ratio of aggregate created dry, harsh environments, which limited the survival of blown-in seeds. The low level of organic material was enough to provide essential nutrients for establishing plants. This 80/20 ratio was adopted and popularized by the FLL in Germany, beginning in the mid-1970s; however, we have since learned that this mix ratio does not hold up across every climate or type of planting and that growing media with greater organic volume may not lose volume as rapidly as predicted. Therefore, green roof builders now use many different mixes with varying levels of organic matter.

Inorganic materials are mineral-based, typically in the form of an aggregate or expanded clay, shale, and slate. Organic materials typically consist of decaying plant matter and stable, well-composted humus.

Another form of organic matter—one that is showing promise in the green roof environment—is *biochar*. Biochar is created by burning biomass under high-temperature and low-oxygen conditions. It is a sustainable byproduct that is lightweight and porous. When included in growing media, it reduces the run-off of nutrients and water. Biochar can also stimulate microorganism populations and has proven to be very stable. The biggest challenges it presents are due to its inconsistent

Fig. 7.12: *A typical aggregate-based growing media with expanded shale ten years after installation.*
Photo Credit: Restoration Gardens, Inc.

properties (because it's produced from many different source materials and varying processes). Because of this inconsistency, it is difficult to determine how much should be added by volume to a growing medium. Some studies have seen biochar increase the water retention of growing media and drought tolerance of plants. In these studies, biochar was incorporated up to 40% by volume.[8] Accessing that sort of quantity would be a challenge for the DIY builder. In fact, biochar is not yet easy to come by, even via the internet. When used, biochar should be *included with* organic matter rather than being a replacement for it.

The organic composition percentage of the green roof will naturally fluctuate. While the initial inputs will break down and decompose, new organics will be recycled into the roof. This is achieved by microorganisms breaking down the debris and leaves that are shed onto the roof. Microorganisms are introduced to the roof through plant roots, organic growing media, and by any natural objects you may place on the roof, i.e., decaying logs from the ground. If your media is aggregate-based, it may be more sterile than is optimal. You can purchase microorganism inoculates, or you can just try adding a few shovels of at-grade soil. Over the years, microorganism populations will naturally increase on the roof, aiding in the continual recycling of nutrients.

While an inorganic-based media will have to be replenished with supplementary nutrients, it will require less maintenance because fewer species will be able to thrive. The opposite can be said for an organic-based media. In addition, organic-based soil will retain more water and keep the media cooler, but it may also be more prone to wind erosion and compaction.

Components

Figure 7.13 is a table showing recommended materials to use when creating your own mix. Always keep saturated densities in mind. Suppliers of manufactured aggregates may be able to provide you with a saturated density; however, the saturated density of natural aggregates will fluctuate. The numbers given below will give you a good enough guide to go on.

Inorganics are separated by type:

- **Natural aggregates** are rocks that have been mined and crushed or screened. These materials are found throughout North America.
- **Natural lightweight aggregates** are natural rocks that have been mined and crushed; pumice and scoria are examples. These materials are igneous rock formed by volcanic activity; they are typically found on the North American West Coast.
- **Manufactured lightweight aggregates** are mined materials processed at high temperatures (1,000°C–1,200°C). This process of heating the raw material expands it to almost twice its original volume, making it lightweight and well suited for a green roof; however, the heating process is energy-intensive.

The table in Figure 7.13 gives values for traditional mineral soils as well as some recycled products.

If you are provided with a material weight in pounds per cubic foot (lb/ft^3), just divide the number by 12 to obtain the 1-inch weight in a square foot space, then multiply it by the depth of your system.

Environmental Concerns

All mining comes with environmental consequences. When natural land formations

Figure 7.13: Common Inorganic and Organic Growing Media Components and Their Saturated Densities [9,10]

Inorganic	PSF (1")	Organic	PSF (1")
Natural Aggregates		Composted yard waste	6.83
Sand	6–11.4	Sphagnum peat	5.83
Sand and gravel, mixed	9.4	Composted pine bark	4.5
Gravel	8.4–9.9	Composted coconut coir	4.9
Pebbles	9.9	Mushroom compost	Contact supplier
Natural Lightweight Organics		Biochar	4.7
Scoria (lava rock)	4.1	Worm castings	Contact supplier
Pumice	3.3		
Manufactured Lightweight Aggregates			
Expanded aggregates (clay, slate, shale)	1.5–2.0		
Expanded clay (often referred to as LECA)	4.5		
Expanded shale (often branded as Haydite)	4.8		
Perlite	2.5		
Vermiculite	0.51–2		
Other			
Brick (solid with mortar) Crushed brick	9.4		
Crushed clay tiles	varies		
Topsoil	8.9–10.4		
Loam	10		

are altered, habitat and natural hydrology are affected. And of course, mining the material requires large machinery for harvesting and transportation.

Expanded materials are highly valued in the industry, but the expansion process can release particulate matter, gases, and volatile organic compounds. Most processing facilities do employ emission reduction technologies such as scrubbers, but they cannot bring emissions down to zero. If you want to use these processed materials, try to look first for locally sourced materials. Secondly, try to look into their extraction processes and, if possible, choose materials that come with the least amount of environmental damage.

Organic material is more sustainable because it is often derived from renewable resources. You might want to be slightly wary of claims that material is produced "locally." Local can be a relative term. Coconut coir, for example, is derived from coconuts grown in Florida, Hawaii, or Puerto Rico (if it can be traced back to production in the United States). When looking into purchasing composted yard waste, check with the supplier to ensure they have quality control practices that reduce residuals from chemicals (fertilizers and herbicides), as well as plastics.

Sphagnum peat is often cited as a sustainable horticultural growing media component, but peat bogs are some of the world's greatest carbon sinks, and sources are not always harvested sustainably. The industry claims that they can remediate harvested sites and return them to their original carbon sink state within 15 years.[11] However, the dredging of peat can dry up these bogs, making them more prone to fires that release the carbon held within.

In addition to its harvesting challenges, peat-based soils can dry out quickly, and they are light enough that they may be prone to displacement in heavy rains. They can also break down and compact easily. If you are building a very small green roof and a peat-based soil is your only option, amend it with a stable aggregate.

Embodied Carbon

Embodied carbon on a dry density basis:

- Common aggregates such as sand, gravel, and pea gravel: 0.00438 $kgCO_2e/kg$
- Expanded clay: 0.39321 $kgCO_2e/kg$
- Perlite: 0.52 $kgCO_2e/kg$
- Vermiculite: 0.52 $kgCO_2e/kg$
- Brick: 0.21 $kgCO_2e/kg$

At-Grade Soil

Often, homeowners ask me if they can just lift the "dirt" around their property and put it on the roof. This is definitely possible; however, there are a few concerns:

- Existing weed seeds from at-grade soil are competitive in nature and can produce plants with aggressive roots that will likely take over the roof.
- Existing pathogens and diseases that are in at-grade soil will be brought to the roof, causing potential damage to your new plants.
- This soil may compact over time or break down, both of which cause volume loss. This can create an unsuitable depth for your desired plants and result in greater maintenance needs to replenish the soil.
- At-grade soil may retain too much water and thus have a heavier loading capacity than one mixed with lightweight aggregate. If used as-is, you may end up with waterlogged soil, which can cause root rot for some plants.
- There are many fine particles in at-grade soil; these can clog a system.

That aside, you can definitely use it, and it is one way to reduce the carbon footprint of the roof. Strategies to remedy the above problems are as follows:

- Try to determine your sand/silt/clay ratio and amend to achieve more desirable characteristics. You can mix the soil with some lightweight aggregate to reduce compaction rates, provide stable media, and increase its drainage capacity.
- Perform some soil solarization. Spread the soil out in a thin 2-inch layer and cover it with clear plastic for a couple months during the heat of the summer to kill off any weed seeds. This will also help kill some pathogens and diseases. You may not get all the bad stuff out, but this process should reduce it.
- Perform your own saturated weight load tests of your remediated soil so that you are aware of the potential loading. See Sidebar for how to conduct this test.

If your budget is very low but you have the time and the willingness to manage the growing conditions of the green roof over the years, or if you're willing to take a laissez-faire approach, then you can definitely use at-grade soil—either as is, or as a component in your mix. Regardless of the type of growing media you use, always keep it covered prior to installation to keep weed seeds out.

The following case studies highlight projects in which we needed to create our own growing media mixed from local materials. It often takes a lot more work, but can reduce the cost of the job.

Getting Everything on the Roof

There are many ways to get the material onto the roof; how you do it will depend on the size of your site and the access to your roof. Here are some of the possibilities:

- Lift cubic yard bags with a forklift and cut the bags to allow the media to drain. This allows you to spread the material while the weight is on the forks.
- Conveyor belts can bring up loose material.
- Front end loader or bucket lift.
- Bucket brigade.
- Roofing contractor ladder lift, normally used for lifting shingles, with buckets.

Saturated Weight Load Test

It's quite easy to perform your own saturated weight load test, and while it may not be 100% accurate, it will give you a ballpark to work with so you don't underbuild your structure. Following are the steps to performing your own test. I recommend doing three tests, pulling samples from various areas, or mixing your soil batch three times; this way, you can take an average of the results.

Build an open-top box that's 12" × 12" × the depth of your system. This can easily be made out of scrap lumber and some plywood. Drill a few holes around the perimeter of the box and the bottom of the box. Line the inside with plastic sheeting. Weigh your box and mark down the initial weight. Fill it with your custom mix and saturate it with a hose until it starts to form an even pool over the top of the growing media. Let it sit for 24 hours. After 24 hours, poke holes through your drilled spots so that you puncture the plastic and allow the water to drain out. It should flow freely. You essentially just want all the water that is filling the void spaces to be drained; however, some water will remain in your test box. Let it drain for about 15 minutes, and then weigh it. Subtract your initial weight from your final weight, and you have an approximate square footage saturated weight of your growing media.

Case Study: Abbey Gardens, Haliburton, Ontario.

See Photos 16a, b, and c. in the Color Section. This green roof designed by Restoration Gardens, Inc. was part of a larger new build project constructed by Fleming College students in the Sustainable Building Design and Construction Program.

We put a 12-inch strip of "granular A" gravel strip along the back of the green roof where it received water from the large upper metal roof to slow down the drainage into the green roof and prevent erosion.

Custom soil was made from ⅜-inch crushed brick, limestone screenings, and local horse manure. We attempted to solarize the manure but were limited with time, so unfortunately/fortunately, the roof was covered in red clover the next year. This turned out to be quite beautiful and was an attractant for the bees! The students planted three small cedar saplings, and now, eight years later, with no supplemental irrigation, they are 6 feet tall. The initial growing media was installed at 5–7 inches, and eight years later, it sits around 3–6 inches. Not all the species from the wildflower mix grew, but there is still full coverage. The roof was built for 80 psf, as the original intention was to allow goats on the roof.

Note: The cedar trees were an student experiment. In general, 5–7 inches of media isn't deep enough to support large plants. We have investigated the roots a few times and will likely remove the cedars before they get any larger.

Case Study: EcoSuperior Garden Shed, Thunder Bay, Ontario.

This roof was created with materials found at a local nursery; it was small enough that it made sense to use some standard bagged products. The mix consisted of 50% lava rock, 20% vermiculite, 10% coarse sand, 10% compost, and 10% pine bark. Ten years later, it still has full cover and has received zero maintenance.

Case Study: Local Farm, Pickering, Ontario.

This project was done with a mix of site soil (sandy topsoil) and "granular B" gravel. These materials were either being removed for the foundation or added for drainage, so the client wanted to use what was on hand. She did her own saturated weight tests and came up with roughly 50 psf for 4 inches. The roof was planted with a native grass mix and watered upon seeding. However, the owner did not want to maintain the roof and was happy for the grass to green up and die throughout the seasons. Ten years later, the roof cycles between hawkweed/moss and grass/moss.

Fig. 7.14: *Site soil is mixed with "granular B" and placed on the roof with a bucket loader.*

- Manually lift with a pulley, or use a ladder lift to hoist purchased bags up and mix on the roof—this is appropriate for smaller roofs only.
- Some commercial growing mix providers will hoist your mix with an articulating crane right to the roof or install it using a blower truck.

Cost

The price of commercial blends will vary by region and by delivery type. Typically, bulk deliveries are cheaper than yard bags; blends that come in smaller bags are the most expensive. Although prices vary widely, here are some ballpark figures for standard sizes you might see offered: bulk $100–150/cubic yard; cubic yard bag $100–150; and 50 lb. bag $7. Custom or homemade growing mixtures have too many variables to list prices.

Saturated weight loads of commercial blends can run anywhere from 48 lb/ft^3 to 96 lb/ft^3.

Amount Required

You will need to determine your amount of growing media in cubic yards, as that is how it is commonly sold. To calculate cubic yards, you will need to know the depth of your system and the area that is to be greened:

(depth in inches ÷ 12) × ft^2 = volume ft^3

volume ft^3 ÷ 27 ft^3/yd^3 = volume yd^3

Sample calculation for a 4" system on a 20' × 20' roof:

(4 ÷ 12) × 400 ft^2 = 133.33 ft^3 ÷ 27 ft^3/yd^3 = 4.9 or 5 yd^3

Optional Layers
Membrane Protection

A membrane protection layer protects the membrane if the drainage layer is incompatible with the membrane or if there will be heavy foot traffic on the roof. Situations where you may require membrane protection include:

- When gravel drainage is used with an EPDM membrane.
- When an EPDM membrane is used on an accessible semi-intensive or food roof system that has hard poly cup drainage.
- When an EPDM membrane has been mechanically fastened—protection is needed to avoid stressing the points where the fasteners lay.
- When the waterproof membrane is chemically incompatible with the chosen root barrier.

There are products available from manufacturers for the purpose of protecting your waterproof membrane. However, a simple way to protect your membrane is to use a thick layer of felt or an extra layer of filter cloth. You could also consider adding insulation between the membrane and the root barrier. This would provide thermal insulation as well as protection. There are several appropriate types, described just below.

Insulation
Role

Insulation provides thermal protection from temperature fluctuations outside the building. Insulation can be part of your building assembly, or it can be part of the roof assembly. Consult your local building codes for insulation requirements.

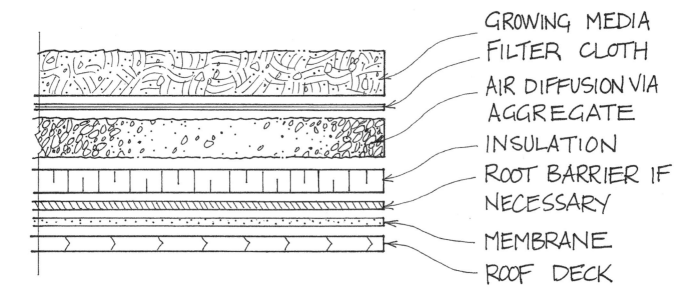

Fig. 7.15: *A typical layered PMR assembly. *You could add another layer of geotextile (filter cloth) below to the aggregate if you are concerned the pea gravel will work its way into the crevices of the insulation or if there is frequent foot traffic on the roof.*

A cold roof assembly locates the insulation just above the ceiling. A soft insulation material, like blanket types or loose fill, is typically used. Between the insulation and the bottom of the roof deck is a ventilated attic space.

Warm roof assemblies locate the roof insulation either just above (protected membrane roof assemblies [PMR]) or just below the roofing material (conventional roof assembly). Warm roof assemblies are most commonly used in flat roof construction; they use a form of rigid insulation that resists compression from the layers above. Materials for conventional roof assemblies can include extruded polystyrene, expanded polystyrene, polyisocyanurate, mineral wool, or perlite.

Most DIY projects that include insulation would consist of a cold roof (not covered in this book) or a PMR assembly. In the case of PMR, the insulation has to be capable of being exposed to moisture.

Green roofs can also be installed over roofs without insulation, such as porches or sheds. However, if you are building over a space that is not insulated and you live in a cold climate, your plants may be susceptible to frost damage. Choose hardy species, including some that can survive in regions with a lower hardiness zone than yours—if they can also survive in your warmer region.

Specifications

When building a green roof with a PMR assembly, lay the root barrier under the insulation and have a layer of air diffusion over top of the insulation. This allows water to evaporate or drain off the insulation; otherwise, it can take on too much water, reducing its thermal resistivity. A layer of pea gravel, pre-formed engineered drainage boards, or other permeable products will provide adequate drainage without impeding open diffusion.

Extruded Polystyrene (XPS)

Extruded polystyrene (XPS) is a roofing insulation that can be exposed to water and can therefore be used in PMR systems. It is usually blue, pink, or green, depending on the brand. This is a petroleum-based product, and most types contain flame retardants.

Embodied Carbon

28.2 kgCO$_2$e/m^2 at RSI 1 (R-5.68)

Thermal Rating

R-5 per inch

Saturated weight per inch thickness

~0.25–0.30 lb/ft^2

Cost of Insulation

Available in a variety of sizes, cost varies by size.

Can range from $1–3/square foot.

Amount Required

The thickness of insulation depends on the desired R-value. The insulation should cover the entire deck.

Water Retention

Water retention mats are common for sloped roofs, areas with little rainfall, or for roofs with plants that require slightly more moisture. A water retention layer will retain water after a rainfall or irrigation and allow it to be taken up by the plants.

These products come in the form of dimpled cups (which can be a part of the drainage system), a retention fleece, hydrophilic mineral wool blankets, or starch-based hydrogels. These need to be placed *above* the root barrier so roots can access the water. But note that water retention isn't always desirable. If you have a flat green roof planted with *Sedum*, for example, having a roof that retains a lot of moisture could actually be detrimental, as sedums are susceptible to root rot.

If saturated weights for the water retention materials are not listed, consult with the manufacturer, or run your own tests, as these items will add weight to the system.

Starch-based hydrogels come in various forms; you may have to order them online or at a hydroponics store. While the gels do store water, their lifespan is short. They also can displace soil or steal water from plant roots after small rain events.

Note that some mineral wool products contain low levels of formaldehyde.

Slope Restraints

Slope restraints provide protection against the loss or movement of material. Loss or movement can result from several factors: there can be surface erosion from weather effects; layers in the system can slip by one another; or the roof angle can be too high for the growing media—exceeding its angle of repose.

Surface erosion occurs when growing media is displaced during strong wind or rain events. This occurs on both low- and high-sloped roofs. This can result in plant roots or underlying layers being exposed. Typically, once the plants have been given time to embed themselves into the growing media, the growing media is relatively stable, and erosion is less of a problem. The following lists offer protection measures at various slopes[12,13].

- For slopes below 10%: A permanent layer of stone mulching, similar in size to pea gravel. This can also retain some moisture on the top level of the growing media.
- Erosion control biodegradable fabric, such as burlap or straw mats, which can prevent erosion until the plant roots have grown enough to hold the soil in place. (Note that while some mat manufacturers say their mats are biodegradable, the material that

Case Study

The slope on this project was 17%. We worked with a welder to create galvanized braces that matched the angle of our ridge. These were placed along the peak every 4ft. and we then attached rot-resistant lumber to these on both sides of the gable. This strategy enabled us to create a steady base in which to fix our horizontal soil restraints.

Fig. 7.16: *Custom corrosion-resistant brackets, designed to match the peak of the roof, are used to hold the slope restraint system in place until roots stabilize the soil.*
Photo Credit: Restoration Gardens, Inc.

breaks down is woven into a plastic net. Sometimes this net is not biodegradable; it can become a trip hazard, and it can also be bothersome for maintenance.) Seams should be joined with twine, tie wire, or zip ties. If you need to hold the sheets down, you can use large rocks or logs. You can also use sod staples, but if you do, be sure to use extra caution to avoid puncturing the underlying layers. When we use sod staples, we cut them to be shorter than our growing media depth and install them on an angle. We also remove them after a couple of years to ensure visitors don't step on them.

- Adding hydrogels, or similar material, which will retain moisture in the system and make the growing media less susceptible to wind erosion.
- Using pre-grown plant matter—in the form of *Sedum* mats—to protect the underlying growing media.

Slipping of layers may occur on roofs starting at about a 10% slope, depending on the design. Here are some ways to avoid slippage:

- Look for combination materials that have factory-welded components. These include drainage sheets with geotextile or membranes with root barriers.
- Avoid layering sheets together that are smooth. Instead, add some friction between by using sheet material with different textures.
- Build supports at the eaves or suspend supports from the peak of the roof. Tying material into the edges can prevent it from slipping.
- Choose growing media components that are highly cohesive.

On roofs with a greater than 17% slope, the ability of the growing media to hold its shape is exceeded, making it susceptible to

Green Roof Material Options   89

sliding. You will need to find ways to hold the material in smaller sections to solve this. This can be in the form of:

- Structural grid boxes made of rot-resistant lumber. Ideally, 4–8 square feet each. These should be structurally tied into the top or bottom of the roof structure to prevent shifting of the boxes themselves.
- A manufactured plastic product in the form of a flexible honeycomb sheet or rigid interlocking grids. These are tied into the top of the roof and filled with substrate;

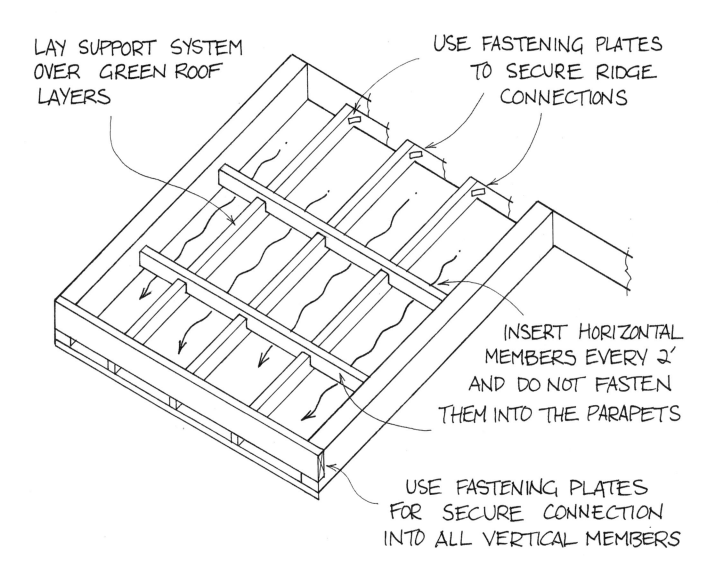

Fig. 7.17: *An example of a sloped roof solution for a small green roof. After installing the root barrier or water retention layer, create a grid system that sits on top of the slope with horizontal members notched above verticals to allow water to pass. Place horizontal members every 2 feet. Drape with filter cloth and fill with growing media. When made out of composite or rot-resistant lumber, these can be exposed to moisture. Note that this is not an engineered approved system. Use on roofs with rafters less than 10 feet and slopes within 10–25%.*

Fig. 7.18: *Plugs planted in an erosion control blanket with a temporary drip system placed on top. Photo 18 in the Color Section shows the roof two years later; you can see how the plants have replaced the erosion blanket.* Photo Credit: Restoration Gardens, Inc.

this system retains the substrate while still allowing water to flow through.
- Tiered slopes with lumber tied into the deck and waterproofed, allowing water to flow through sections.
- Modular units with strapping.

In the Color Section, in Photo 17, you can see how a *Sedum* mat was used to cover a steep peak on a residential roof.

Wind Erosion

In areas with high winds, it may be necessary to cover the green roof with something temporarily to resist wind scour of the growing media. Wind erosion is similar to surface erosion, but it can happen even on flat roofs. Look at employing the same measures given for surface erosion if your region is prone to strong winds or if an engineer has recommended you do so.

Additional solutions to mitigate the damage by the wind include the design/installation of non-vegetative zones around the perimeter of the roof. Instead of plants, you install heavier materials that will resist the wind scour (see "Pathways/Vegetation-free Zones" near the end of this chapter).

Keeping the soil moist is also a wind-erosion control measure; for very windy areas, consider installing a permanent irrigation system. Regions prone to tornadoes or heavy winds will need engineered wind protection; otherwise, building a green roof is not recommended.

Irrigation

There is no such thing as drought-proof plants. All plants need water to grow, especially during the critical establishment period. Plants require a few weeks of daily water once planted, although the amount and frequency will be dependent on the season in which you plant your roof. In addition to the season, water will vary based on the method of plant installation you choose.

After establishment, it is not recommended that you provide irrigation (except in special circumstances). Plants that are regularly irrigated will grow larger but be less tolerant of drought. Food roofs and arid regions are exceptions because small amounts of supplementary water are required to keep the roof alive. In addition, some regions will experience a greater number of days with extreme heat due to climate change; therefore, it may be wise to install a permanent irrigation system. You do not need to run it constantly, but should your region experience above-average drought, it will be in place. It's a small expense to protect your larger investment.

Using potable water to irrigate a green roof is unsustainable in most cases. Try to collect rainwater from another roof on your site. The green roof itself will not provide a lot of run-off, as the nature of its design is to retain most of it. A small pump can help you get the water onto the roof.

There are a couple ways you can irrigate your roof:

- **Temporary Overhead Water**—either by watering yourself with a hose and nozzle or by placing a sprinkler in the middle of the roof. You can attach hoses to at-grade hose bibs or install one on the green roof level. The higher the roof, the more pressure that is required from grade to push water up the hose. As to amount, I often recommend that owners water the area evenly and stop once they see small puddles forming on the surface; this ensures a deep soak that gets water to the roots.
- **Drip Irrigation or Soaker Hoses** that deliver water to the roots. These components should be located *in* the soil, rather than on the surface, to prevent rapid evaporation. This will also favor the roots of your intended species rather than seeds that blow in. You will have to space your lines closer than you would in at-grade gardens because most green roof media is aggregate-based. Aggregate soils will drain faster than at-grade soil, losing out on horizontal wicking through capillary action. We normally space lines about 12 inches apart and put them in about 1 inch below the surface.

The advantages of drip irrigation are that less water will be lost to evaporation or run-off and that larger areas can be watered with less pressure.

If you need or plan on having irrigation, plant a green roof with plants that have similar watering requirements, as this will simplify your watering needs. The exception is a sloped roof, as discussed earlier.

Details

Edging, Flashing, and Drip Edges

Edging is a barrier between the growing media and any non-vegetated zones that allows water to flow through it or under it toward the drains. Edging can be plastic, rot-resistant lumber, or corrosion-resistant metal. Check in your region if combustible materials (such as traditional lumber) can be used on your green roof. It's easier to use a material with an L-shape so it will stand on its own while you fill it with soil—something similar to what landscapers would use when laying brick pavers. If placed on the drainage layer, it does not need to have perforations for water, as water will be diverted under it.

Flashings cap the parapets, protect penetrations, or cover terminations at adjoining walls. These would be designed the same for traditional flat roofs. Flashing for wooden parapets and at walls should be made of corrosion-resistant metal. Brick parapets can also be topped with a coping stone. The area around the roof perimeters and corners receive the highest wind pressures, so it's imperative that these are secured tightly with the proper fasteners. Vent penetrations are often flashed with PVC, rubber, or plastic; sometimes, metal boots are used. Ensure that the flashing material that you use is compatible with your membrane. The membrane manufacturer will list and/or sell such compatible products.

Fig. 7.19: *An example of an edging product surrounds some installed modular trays as well as a non-vegetation zone around a plumbing stack.*

Photo Credit: Restoration Gardens, Inc.

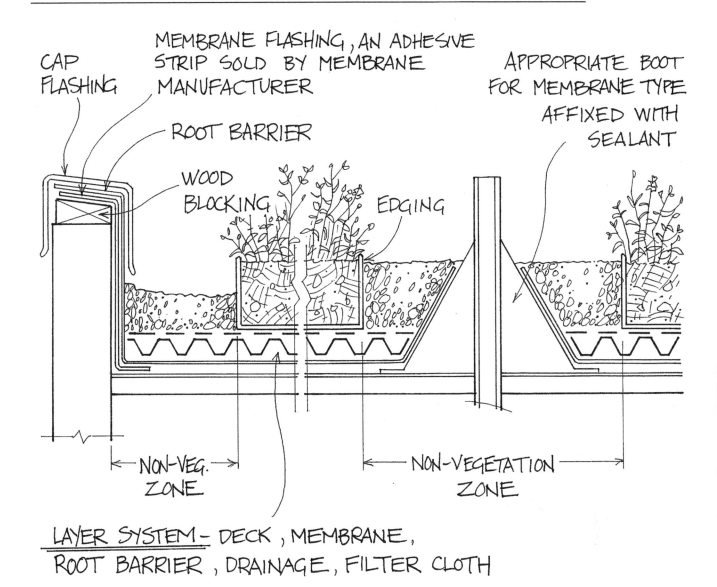

Fig. 7.20: *Flashing for typical parapet walls and a typical roof penetration.*

Roof penetrations and the flashing around them can be vulnerable areas on a green roof. Reduce risks by limiting or eliminating penetrations, using appropriate caulking and seam material, and choosing a membrane that is easier to work with around penetrations, such as a liquid-applied product. You should inspect flashing after major storm events or at least twice a year.

Drip edges direct the flow of water down and off the fascia board of the roof. Drip edges can be made of corrosion-resistant metal or composite lumber. Always make sure any material that is being connected/adhered to your membrane is a compatible material.

You will see examples of how to handle a drip edge in Chapter 8.

The three commonly used types of corrosion-resistant metal on green roofs are:

- **Aluminum:** Aluminum isn't as strong as steel. It does not corrode and is often sold in a variety of colors. Aluminum products are often high in recycled content and

can be manufactured with high-quality baked-on resin finishes that meet the National Sanitation Foundation (NSF) standards for rainwater harvest.
- **Steel:** Galvanized steel will resist rust, but it will leach zinc into the run-off,[14] so opt for powder-coated materials, if possible.
- **Copper:** Copper is a sturdy metal with a classic look, but it will leach into the water.

Pathways/Vegetation-free Zones

Pathways or vegetation-free areas are designed into a green roof for maintenance personnel, additional drainage, as places where plants will not grow, and/or for areas that may be more susceptible to leaks or wind uplift. These areas are typically around the edges of the roof (where it meets a vertical surface), around roof penetrations and mechanicals, or under an adjacent roofline that either blocks the rain or drains excessive water onto the green roof.

These areas can be made with pavers, gravel, or washed stone. In part, stone used to separate the vegetation from the perimeter of the roof serves as a firebreak in the event of a fire on the roof. If using gravel, use at least 3 inches to prevent wind uplift.[15] Check with your municipal code to determine what fire prevention design strategies are required. In Toronto, anything constructed or clad in combustible materials must have a vegetation-free border of a minimum of 1.5 feet (0.5 meters) or equal to the vegetation height at maturity (unless approved otherwise[16]).

I highly recommend laying these stone materials overtop of the membrane, root barrier, drainage, and filter cloth to protect the membrane from UV as well as abrasion from footsteps on the pathways.

Efficient Design

Take some time to review your plans. See if you can find materials that serve more than one purpose. These will save you time and sometimes money. Some examples of these include:

- A high-density polypropylene dimple board can act as a root barrier and drainage on low-slope extensive roofs if seams are welded.
- Insulation provides thermal barriers, can act as a protection board, and can add topography features under the growing media without adding weight.
- Hydrogels offer wind erosion protection as well as water retention.
- Pea gravel can be used as the drainage layer as well as surface erosion protection and water retention.
- Biochar can increase water and fertility availability in the organic content as well as provide a sustainable, lightweight, stable media.
- Natural features such as logs can be used in some designs to prevent surface erosion, hold erosion sheet material down, and to create microclimates and habitats for insects.
- Lightweight aggregates can be used in the growing media as well in some vegetation-free areas.
- Some commercially available products serve as protection boards, drainage, and root barriers. These "sandwich" boards are easy to install, but the water in the drainage layer is not accessible to the plants. If you have a flat (0–2%) roof slope, these boards might be suitable for extra drainage.

Modular Units

There are lots of modular options available for purchase. Some you plant yourself, and some modules come pre-planted. The challenge will be finding products that work with your budget. Pre-grown modules come with a higher price tag, as the nursery has done the work to plant and care for these modules for over a year. This price can sometimes be offset by reduced labor time on site, but if you are doing the labor, this may not be to your advantage. Pre-planted modules are heavy, and they usually have to be shipped, which adds to the cost.

Modular units can be beneficial in situations where access to the roof is a challenge, where slopes are present (because modules naturally stack beside each other), or if time is an issue. Drip irrigations systems are a challenge with modular trays. Irrigation needs to be provided via a temporary overhead method, or you have to put in a permanent overhead spray system. If the trays get damaged, you have to ensure hard plastic edges do not come in contact with underlying layers. In addition, cutting trays to fit around penetrations or curves can be tricky. It can be done, but the job will require power tools, and you will inevitably waste some of the materials.

Please always discuss requirements for the modules with the growers/suppliers in terms of ideal membranes and slopes as well as installation or accessory requirements.

Ordering Materials

When you are calculating the amount of material you require, refer back to your measurements and account for overlaps and waste. A simple solution is to always add 10% to the amount of material. However, most materials have to be ordered in minimum quantities or order sizes. If that's the case, always choose the larger amount rather than the smaller amount—unless you redesign your layout.

Fig. 7.21: *A fully mature modular tray by LiveRoof Inc. See Figure 2.2 and 2.3 in the color section to view the diversity of plants and see the modular trays on a sloped roof.*
Photo credit: Kees Govers

Fig. 7.22: *A sloped install built with 4″ LiveRoof modular trays.*
Photo credit: Kees Govers.

Chapter 8

A Rural New Build

In this chapter and in Chapter 9, I present step-by-step accounts of real-life projects. This chapter gives the steps we went through on a project for The Municipality of Centre Hastings, contracted by Fleming College's Sustainable Building Design and Construction program. Natvik Design, Inc. was subcontracted to teach the students how to build a green roof. This green roof was on a shed roof overtop an open porch in a rural setting. The example in Chapter 9 is a flat roof with parapets on an urban site. These step-by-step procedures should serve as a reference for the installation of your own project.

Case Study

The Project

On a rural build, a covered porch was designed to accommodate a green roof. A mix of sedums and hardy native perennials were specified, as the porch was open underneath with no insulation. This project was designed by Natvik Design, Inc. and installed by building students.

Roof Specs:

Size: 600 square feet

Slope: 2%

Design Load: The roof could accommodate 32 psf (1.5 kPa) of green roof (not including the roofing assembly).

System Depth: 4 inches

Roof Type: Shed roof

Material was brought up via scaffolding and manual hoisting.

Fig. 8.1: *The completed green roof over the open front porch.*
Photo credit: Chris Magwood

Layer Construction

See Figure 8.2 for a profile of our build-up.

Tools:

- Circular saw and miter saw with blades for standard carpentry
- Measuring tape
- Speed square
- Drill
- Drill bits and drivers
- Nut driver
- Hammer
- Hard rake
- Marker
- Blade

Materials:

- Parapet: 2 × 6 dimensional lumber
- Soil restraint edge: Cedar 2 × 6
- Corrosion-resistant metal brackets
- Membrane: EPDM root-resistant membrane, loose-laid
- Aluminum drip edge
- Ice and water guard
- Separation sheet/supplemental drainage: Platon
- Drainage: drainage cup panels
- Filter cloth: Terrafix 360r
- Growing media: custom mix of crushed brick and duck compost
- Fascia: 1 × 8 cedar
- Flashing: aluminum pre-formed flashing
- Galvanized or coated roofing nails
- Exterior coated screws
- Galvanized roofing screws

What We Did

Roof Prep

1. Built up the side parapets to contain the depth of the system. In this case, we started with two layers of rough sawn 2 × 6. We would build another 4 inches once the membrane was in place.

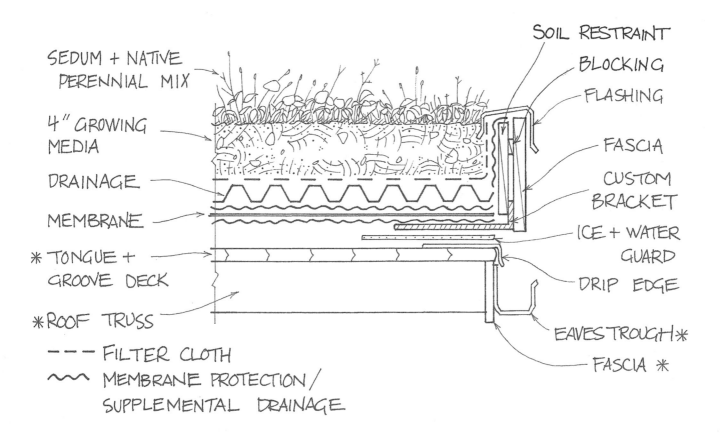

Fig. 8.2: *The build-up for this porch green roof. *Note that some elements are not included in the step-by-step account given here.*

Fig. 8.3: *Roof details showing initial parapet height with drip edge flashing covered by the ice and water guard. Our metal brackets will sit an inch off the eaves of the roof and fasten above each truss.*
PHOTO CREDIT: LESLIE DOYLE

2. At the end of the joists, we installed the 1×8-inch cedar fascia boards.
3. Any areas on the deck that appeared rough or out of plumb with the rest of the boards were sanded down, and we ensured all fasteners were sunk or flush to the deck.
4. We swept the deck to remove all debris.

Waterproofing

1. Installed drip edges on the edge of the eaves using galvanized roofing nails spaced 12 inches apart.
2. Rolled out and adhered ice and water guard to be flush with the edge of the deck, over the drip edge flashing and fasteners. This prevents water from getting into a roof when there is ice damming in the winter. This is necessary on the eaves when loose-laying a membrane—unless you drape the membrane over or adhere it to the edge.
3. As our roof had an open drain edge, we used custom-made brackets (any metal fabricator can make these) to hold up our eaves' retention profile. The face of the bracket overhangs the roof to allow for water to drain behind the fascia board once it is attached to the soil restraint edge. The brackets were nailed in above the self-sealing ice and water guard to be covered by the membrane. We installed a bracket over each truss.
4. As we were using EPDM over bitumen ice and water guard, we needed a separation sheet. On this project, we had some extra dimple board on site, so we laid it with the dimples down to provide a

Fig. 8.4: *After the membrane was draped in place, we fastened it with two more layers of our parapet to prevent tension on fastener points. Our water-shedding root barrier and flashing will cover these additional parapets.*

Photo Credit: Leslie Doyle

flatter surface; this also protected the membrane from snags in the rough sawn lumber. The dimple board is tacked into place on top of the parapets with galvanized roofing nails.

5. The EPDM was laid down and draped over the edges, making sure it sat tightly in corners. The best way to lay out EPDM is to bring the measured, pre-cut membrane up to the roof *folded*. This way, it can be gently unfolded into place without having to drag it over the roof. Make it easy on yourself and cut the EPDM somewhat larger than it needs to be, then trim it flush once it's held in place.

6. We held the EPDM in place on the outside of the parapet with strapping for the side fascia boards followed by two rows of 2 × 6 for the remaining parapets. This way, when people walked on it for the remainder of the install, it didn't get pulled at the fastener points. Water will not be able to enter this joint, as we would provide more coverage later. At the back of the roof, the EPDM was fastened to the ply facing.

Open Drain Edge

1. Using shims, the soil restraint was fastened onto the back of the custom brackets, high enough to allow water to flow through the layers and off the roof. The finished height met the height of the side parapets.

Fig. 8.5: *After the membrane protection was installed, we used temporary shims to lift the soil restraint to allow for drainage. The height will retain 4 inches of growing media and match our side parapets.* Photo Credit: Leslie Doyle

Essential Layers

1. Dimple board provided further drainage, and a root barrier (which was optional, in this case) was then installed in a water-shedding pattern above the membrane and tacked into the top of the parapet.

2. Drainage boards were snapped into place above the root barrier.

3. We cut a small piece of dimple board and tacked it to the inside face of the retention edge to prevent it from sitting in water. This would not be necessary if you used composite lumber. Water can still percolate under this piece.

Fig. 8.6: *This project used rigid drainage trays, which allow for some slight water retention, but most water will drain through and under the cups toward our open edge.*

PHOTO CREDIT: LESLIE DOYLE

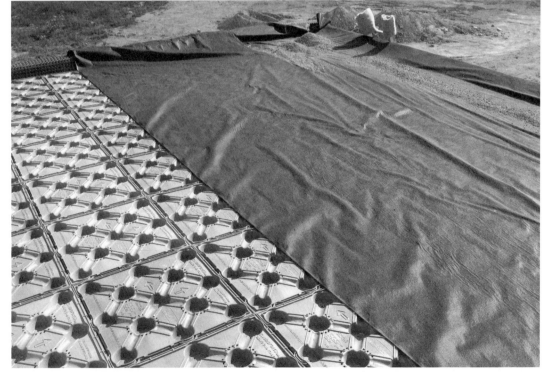

Fig. 8.7: *A shot of some layers. We installed some growing media to prevent the filter cloth from billowing up as we finished the rest of the coverage.*

PHOTO CREDIT: LESLIE DOYLE

Fig. 8.8: *With many hands, the growing media gets lifted in no time!*

Photo Credit: Leslie Doyle

4. Filter cloth was laid overtop and fastened into place on the top of all the parapets and retention edge.
5. Growing media was mixed on the ground and hoisted via many hands in 5-gallon buckets. A bucket brigade!
6. A 16-square-foot wooden frame made of 2 × 4s was used to facilitate filling in the media and ensure even coverage.

Plants

1. Hardy plant plugs of sedums and native perennials were distributed around the roof with a spacing of 1–2 per square foot and planted immediately. Plugs were planted deeply enough to ensure the entire root ball was covered, and the growing media was gently packed around the plant so water would not displace the surrounding media.
2. Plugs were watered upon completion.

Finishing Details

1. To complete the roof, we used blocking to face the drain edge with a 1 × 8 fascia.
2. We then installed pre-fabricated aluminum flashing over the soil restraint and the parapets using galvanized roofing screws. These should be caulked or replaced in 10 years, as the rubber can become brittle over the years.

Open Drainage Alternative

There are lots of ways to finish an open drain edge. One that I have often used is from Clarke Snell and Tim Callahan's book *Building Green: A Complete How-To Guide to Alternative Building Methods*. Figure 8.9 is a sketch of it. In this approach, expanded metal grates are used with composite lumber as a drip edge. The benefit of the composite lumber is that it outlasts natural wood and reduces the need for an aluminum drip

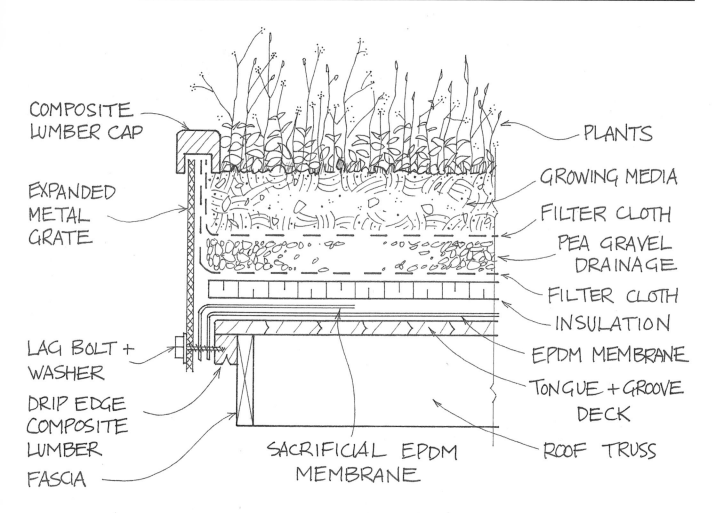

Fig. 8.9: A sketch showing an open drain edge alternative, adapted with permission from Snell and Callahan, Building Green. Green roof layers differ slightly from those in their book.

edge. One detail I liked was that they routered their 2 × 6 deck boards along the edge to make a curved edge to resist abrasion with the membrane. Note the "sacrificial membrane" piece that protects the main membrane from UV.

I have also used this method with cedar fascia and cap and an aluminum drip edge. See Figure 8.10 for the completed look. I used ¾-inch 9-gauge, flat carbon steel that was galvanized. The diamonds were ¾ inch by 1½ inch. The sheets were 4-foot by 8-foot, and we had them sheared to 6-inch widths, giving me 8 pieces that were 12 feet long. Overlap the seams by a few inches. Note that the grate was galvanized and then sheared to length, so the tips will rust. Plus: Be careful. The live edges are extremely sharp. I have the scar to prove it!

Additional general installation tips are listed at the end of Chapter 9.

Fig. 8.10: *This roof has a grate edge with cedar fascia and cap and an aluminum drip edge. See Photo 19 in the Color Section for a roof we completed with materials given in the Snell and Callahan example.*
PHOTO CREDIT: RESTORATION GARDENS, INC.

Chapter 9

An Urban Retrofit Build

Case Study

The Project

In this urban build, a replacement membrane was required, so a green roof was planned and installed by Restoration Gardens with a roofing contractor installing an EPDM membrane. The system is built as a PMR, but with the layer of insulation incorporated to enhance interior ceiling insulation. One internal drain was covered with a drain box. The scupper was surrounded by clear stone and gravel restraint.

Roof Specs:

Size: 225 square feet

Slope: 2%

Design Load: The roof could accommodate 50 psf of green roof.

System Depth: 4 inches

Roof Type: Flat roof with parapets

Material was brought up through the house and out a patio door.

Fig. 9.1: *An urban build on an existing roof with a mix of* Sedum *and native plants. Shown here are* Sedum kamtschaticum *and* Penstemon hirsutus *in bloom.*
PHOTO CREDIT: RESTORATION GARDENS, INC.

Layer Construction

Figure 9.2 shows a schematic profile of the green roof build-up.

Tools:
- Circular saw and miter saw with blades for standard carpentry
- Measuring tape
- Speed square
- Drill
- Drill bits and drivers
- Nut driver
- Metal grinder
- Hammer
- Hard rake
- Marker
- Blade

Materials:
- Parapet: 2 × 6 dimensional lumber
- Membrane: EPDM membrane assembly—supplied and installed by roofing contractor
- Drainage: Platon
- Insulation: DOW Roofmate insulation
- Drainage/Air diffusion: Drainage board by ThermaGreen (with filter cloth attached)
- Filter cloth: Terrafix 360R for around drains
- Soil restraint: GeoEdge by Permaloc
- Scupper/Drain boxes: GeoEdge drain boxes by Permaloc
- Growing media: Engineered growing media from Gro-Bark supplied by blower truck
- Galvanized or coated roofing nails
- Exterior coated screws
- Plant plugs
- Garden hose and oscillating sprinkler

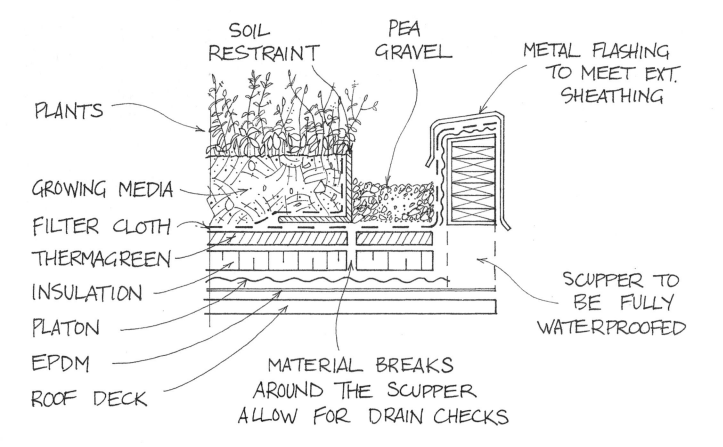

Fig. 9.2: *Layer construction for this urban extensive roof system.*

What We Did
Roof Prep

1. This roof required a new membrane, so the previous tar and gravel roof was removed, which allowed the contractor to check for structural damage. New parapets, 12 inches high, were stacked from 2 × 6 lumber and fastened to the deck. The system was designed to have one internal drain to be covered by a drain box, and one scupper 6 inches wide by 4 inches high, which was cut into the perimeter of the parapet.
2. Roofing contractors were called in to install a new, fully adhered EPDM membrane.

Fig. 9.3: *New EPDM membrane installed.*
PHOTO CREDIT: RESTORATION GARDENS, INC.

Green Roof Layers

1. Because the slope was very low (< 2%), dimple board was used to allow water to move under the insulation; the dimple board was installed in a water-shedding pattern starting on the scupper edge. We brought this material up and over the membrane to protect it from UV damage, cutting away material at the drains.
2. Three-inch DOW Roofmate shiplap insulation was added for additional insulation on the house.
3. Additional drainage and aeration was added using material made out of recycled rubber with an attached filter cloth by ThermaGreen. This was placed over the entire insulation surface.
4. Additional filter cloth strips are used in front of drains.

Fig. 9.4: *Dimple board installed to move water under the insulation and protect the membrane from UV.*
Photo Credit: Restoration Gardens, Inc.

Fig. 9.5: *Installing the insulation in an alternate seam pattern.*
Photo Credit: Restoration Gardens, Inc.

Vegetation-Free Zones

Vegetation-free areas were created with edging and pea gravel in front of the overflow scupper and a drainage chamber was placed over the internal drain. Metal grinders can be used to cut edging pieces to length. You can use small offcut pieces and self-tapping screws to hold two butt ends together, or some products come with attachment pieces to prevent ends from shifting. You can place cut strips of filter cloth inside the edging to prevent growing media from washing into the vegetation-free areas.

Growing Media and Plants

1. As there was no place for growing media delivery on this small urban lot, a local soil blowing company was hired to blow in the engineered growing media.

2. Plug plants were distributed around the roof with a spacing of 1–2 per square foot and planted immediately. Plugs were planted deeply enough to ensure the entire root ball was covered, and the growing media was gently packed around the plant so water would not displace the surrounding media.

3. Plugs were watered once planted.

Fig. 9.6: *Vegetation-free zones are installed. The internal drain is covered by a drainage box with a lid.*
PHOTO CREDIT: RESTORATION GARDENS, INC.

General Installation Tips
Weather

Some poly material is affected by changing temperatures. If you are installing in the warmer afternoon, it may stretch—only to contract during the cooler nighttime temperatures. So try not to pull material too tight when fastening.

If only some of the components of a green roof are to be installed before first frost, we install all the layers up to the filter cloth. Temporary weights are placed along the seams and around the edges to protect the filter cloth and the assembly below it from winter winds. In the spring, before installing the growing media, we sweep it clean, shake it off, or even do a light shop vac. If you have a loose-laid membrane or a membrane that requires the weight of the growing media, you must install the growing media when you install the other components. Cover it with an additional layer of filter cloth to prevent wind erosion and weed seeds, as noted in Chapter 7.

Caution should be used with any green roof material on windy days. If you are using a blower truck for growing media, the light particles (organics) can blow away into a neighbors window (ask neighbors to close windows) or just blow around the roof. Sometimes the soil provider will wet the material before putting it into the truck. You can also do a light spray once the material is on the roof, but don't oversaturate it because it will stick to your boots as you finish the install.

Parapet Construction

If your municipality has parapet height requirements, build to accommodate them. If there are no prescribed heights, parapets should be built high enough to account for the total depth of the green roof, including the thickness of all the layers. Add 2 inches above the final grade of the growing media to prevent growing material from surface erosion displacement.

Create an angle on the top of the parapet so cap flashing can drain onto your roof.

Layered Material

For layering materials with overlaps, start at the bottom of the slope and work your way up. This is important. You want the water to shed off the roof rather than under your layers.

Ensure layered material is fastened on top of a parapet, or high enough on the wall, and is covered with appropriate flashing. Use galvanized or coated roofing nails to secure materials.

For retrofits, there are a lot of tricky cuts around penetrations, corners, and upturns. If you cannot avoid seams, ensure you use appropriate seam accessories for your membrane type.

Use cant strips or ensure there is enough slack for inside wall connections because material will pull once the roof is filled with gravel and growing media.

To place the root barrier over a vent stack, fix the root barrier to a parapet and lay it into position. When you get to the penetration, mark on the root barrier where the penetration is. Then lift it off and cut an X so that the root barrier can slip over the penetration. You could also measure out the center point of the penetration onto the root barrier and then cut an X. This can take some practice to get it right—which is another reason why vegetation-free zones are often installed around penetrations; they are weak points in the root barrier.

There are some terrific videos online that show you how to cut layered materials for covering inside corners. There is an art to getting sheeting material in place neatly without bulk. Practice a couple times before making the final cuts. Figure 9.7 provides some guidance on a basic pig ear bend.

If the flashing is already in place on parapets, use a putty knife to help you slip the root barrier under the flashing. Trim any excess to avoid it buckling and bulging the flashing.

Growing Media

An easy way to ensure level grading as you move along is to tape or mark the desired depth of your system on your hard rake. As you rake out the growing media, turn the rake upside down to check your depth. Always fill slightly higher, as the material will compact with the first watering.

Never blow or dump a bulk pile of growing media on the roof, as this can overweigh the roof's loading capacity.

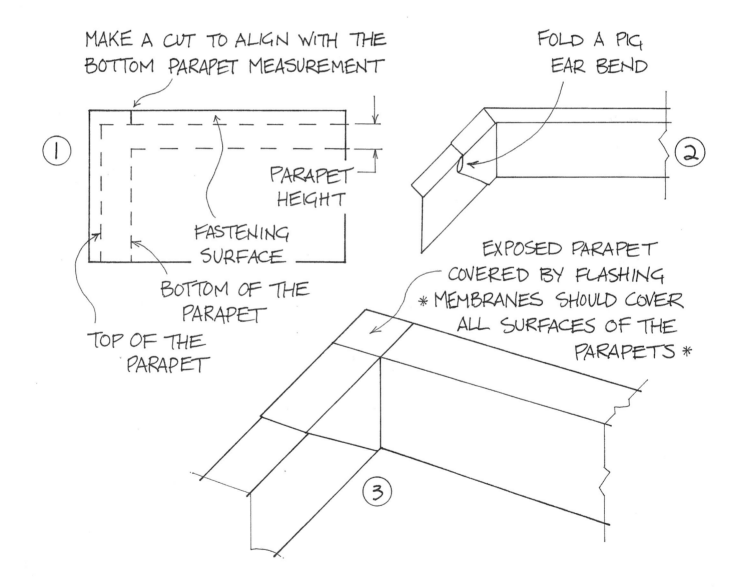

Fig. 9.7: A pig ear bend to get sheeting material tight against parapets. Start by measuring and marking out the height of the parapets on the sheeting material. Then make a small cut where indicated.

Chapter 10

Maintenance

Your attention to the green roof should not stop once the plants are established. You have now created a living system, and it deserves some attention every year—if only just a couple visits. Fertility changes, extreme weather events, and natural competition can all change the soil and plant dynamics. Maintenance should not feel like work; instead, take this time to enjoy your green roof. From the rooftop, you can enjoy new vistas, gain an understanding of how the green roof ecology evolves from season to season or year to year, and look for evidence of visitors. It is during maintenance that I get to see just how many bees and spiders enjoy the new green space, and I always love the surprise of discovering some eggs!

Maintenance is required more frequently during the establishment period and the first five years of the green roof than in the long term. It is during this time that efforts need to be made to ensure your plants have the water, space, and nutrition they need to create a healthy root system and develop resiliency. Creating a healthy green roof from the beginning will result in less maintenance over time.

The following maintenance recommendations are for those who wish to maintain the initial planting plan as closely as possible. For those of you who wish to let nature run its course, the following information may be of interest to you, but your priority will be to follow the establishment suggestions found in Chapter 6. Some regional municipalities may require a maintenance plan for a green roof. This may be in the form of a 3–5 year plan listing annual activities. As well, some insurance companies may require a certain level of care for continued coverage. Check with these sources to ensure you are meeting basic requirements.

Fig. 10.1: *A duck egg is found sheltered on this roof located in a park.*
Photo Credit: Restoration Gardens, Inc.

Irrigation

During the first three weeks after planting, water is required daily—or twice daily, if seeds and/or cuttings are used. Water can be in the form of irrigation or rainfall. Irrigating the roof is best done in the early morning to reduce evaporation. I suggest making it part of your morning coffee routine.

After three weeks, you can reduce the watering to 2–3 times a week. You want to wean plants off of their steady water diet to encourage drought tolerance. Tapering off will reduce the stress on the plants. Your goal should be to get the plants established to the point where you only have to water during hot summer months or droughts.

For watering during that first crucial year, I advise clients to give the roof a good, thorough watering—which means watering until puddles form on the growing media or a significant amount of water begins to drain off the roof. I say "significant" because, depending on the design of your irrigation system, some water sources may be located next to the drain and begin to drain almost immediately. Try to avoid this type of design, if possible, to reduce water consumption.

After the first year, your green roof should not require water unless your region experiences unseasonal drought or a heat wave. If your plants are suffering during a drought or heat wave, get water on the roof. If there is a major dieback on the roof, try to plant some more resilient species; otherwise, these gaps will fill in with blow-ins.

Nutrition

In addition to the sun's energy and the hydration of water, plants require a range of nutrients to grow. Your initial organic content will be sufficient for the first year but will likely need supplemental fertilization for years 2–5—or until a natural recycling system forms on the roof. Regions with high heat and humidity experience a faster decrease of organic matter than do cooler and drier regions.

The percentage of organic material you add to your growing mix will be based on the types of plants that you want to grow, as well as other factors discussed earlier, especially in Chapter 6. Choosing green roof plants that are more stress tolerant than others can reduce the amount of supplemental fertilization that is required. In addition, choosing a component such as biochar can retain nutrients on the roof for longer periods of time. Good green roof design should reduce the reliance on long-term resource needs.

There are some sustainable organic amendments that can be added to the roof. The choice may be based on the ease of acquisition as well as the ease of getting them on your roof. If you can walk up to your roof or access it through a window, heavier materials may be an option for you. However, if you must climb a ladder, something light is more practical. These amendments include:

- Liquid kelp extract
- Fish emulsion
- Compost tea
- Worm castings
- Compost

These products can be applied to the soil in the spring and again in the fall, if desired. Liquid products can be applied using a backpack sprayer, a small pump sprayer, or even a watering can. Compost is considered a garden miracle product as it improves the soil structure, increases aeration and water retention, and benefits soil microorganisms.

You may notice that the run-off in the first year is a bit brown; this is due to excess nutrients and humic acids in the initial media.

Don't worry about this. The nutrients are not at harmful levels, and the brown run-off will taper off after a few months.

An alternative to the natural products just listed is a commercial coated slow-release fertilizer. While it may not produce lush biomass shortly after application, slow-release fertilizers will supply the plants with nutrition throughout the growing season, resulting in stronger plants the following year. This type of fertilizer also reduces nutrients leaching into the run-off.

When looking for commercial fertilizer, you will be presented with three numbers on the box. These represent the percentage in weight of available nitrogen, phosphorus, and potassium, respectively. Nitrogen is necessary for leaf growth; phosphorus for healthy roots, flowers, and fruit; potassium encourages growth and healthy plant cells. A balanced fertilizer (e.g. 5–5–5 or 14–14–14) will be sufficient for a green roof environment. Follow the instructions on the box or be a little more conservative than the box suggests. Too much fertilizer can damage your plants and leach nutrients into nearby water systems. Check with your local environmental authorities because some phosphorus-containing fertilizers may not be allowed in some watersheds. Always water the garden after you apply supplemental nutrients, and never apply when it's hot and dry, as fertilizer can burn leaves.

If your plants are showing significant dieback and signs of damage, and water does not seem to help, consider talking to a local university or soil lab about getting your soil tested for nutrient deficiencies.

Weed Control

The term *weed* is subjective. Some plants on the green roof list may be considered a weed in your at-grade garden. On the roof, a weed is any unwanted plant that competes for space, nutrition, and water. Weeds enter the roof via wind, birds, and animals—or you. If you weed your garden and then climb up onto the roof, your boots or tools may contain soil with weed seeds. In addition, depending on how sterile your initial growing media components were, weeds may have been brought in at the beginning of your project.

The best preventative strategy for controlling weeds on green roofs is establishing conditions that do not favor weed growth. The following cultural practices should be followed to avoid creating habitats suitable for weeds:

- Monitor and improve soil conditions if necessary, i.e., improve compaction, drainage, or nutrient levels.
- Decrease competition by encouraging quick establishment, employing dense planting plans, filling in gaps, or using species that spread.
- Maximize green roof plant health by providing adequate water and nutrition suitable for your plant species.
- Always maintain your roof with clean tools and boots to prevent the introduction of new seeds.

Some green roof owners take a laissez-faire approach to their green roof. Once it's planted, they are happy to let nature take over. There is nothing wrong with this approach as long as you understand that there is a chance you may end up with one dominant plant species—a species that may thrive and then die out in the heat, leaving you with periods of brownouts, or worse case, an empty roof. However, you *could* end up with a roof with beautiful diversity and ever-changing aesthetics.

If you prefer a more pro-active approach, I recommend visiting the roof 3–5 times a year for the first couple of years. The frequency will depend on your region, the length of your growing season, and when you planted the roof. The idea is to walk the roof and pull any weeds out before they have a chance to set seed. This is important, as heavy weed infestations can be hard to eradicate. Once you start to notice weeds, try to identify the species to determine if they are a nuisance or just an aesthetic problem. Once a weed is identified, you can learn about its lifecycle to determine the most effective and efficient method of removal.

The following species are ones that we found were very important to remove (in our US zone 6a/Canada zone 6b). Once you have a large infestation, these plants are a challenge to eradicate. These species either produce thousands of seeds or create spreading webs of rhizomes; the webs can weave around desirable species, making it a challenge to pull out just the plants you don't want.

- *Medicago lupulina,* black medick
- *Cerastium vulgatum,* mouse-eared chickweed
- *Trifolium pretense* and *T. repens,* clover, red and white
- *Polygonum aviculare,* prostrate knotweed
- *Galium mollugo,* smooth bedstraw
- *Chamaesyce nutals* (syn. *Euphorbia nutans*), nodding spurge
- *Chamaesyce maculate,* spotted spurge
- *Vicia cracca,* tufted vetch
- *Digitaria sanguinalis,* large crabgrass
- *Digitaria ischaemum,* smooth crabgrass
- *Panicum dichotomiflorum,* fall panicum
- *Panicum capillare,* witchgrass

Fig. 10.2: *When vetch appears on the roof, it needs to be removed as soon as possible. Here, it was left too long, so it set seed and spread. Now it's crowding out desired* Sedum *species.*
Photo Credit: Restoration Gardens, Inc.

Keep in mind that aggressive weeds and unwanted trees should always be removed throughout the life of a green roof.

In Ontario, the use of most pesticides and herbicides for cosmetic purposes is prohibited. Those that are permitted in home gardens and lawns must be specially requested, as they are not readily available on store shelves. However, these chemicals are never recommended for green roofs because they can negatively affect the roofing membrane (and other roofing components) or flow into run-off. As a result, manual, mechanical, and organic weed control methods must be utilized to manage weeds effectively on green roofs.

When weeding—especially if the roof is relatively small—it's best to work with small hand tools to avoid damaging underlying layers. With hand tools, you can ensure you get the full root out. When pulling up the weeds, do your best to shake off any loose growing media; you want to hold on to the limited amount that exists on the roof. Pull all the weeds from the non-vegetated areas as well. If you've put these areas in, you may as well keep them free and clear. Any seeds produced by plants in these areas will inevitably make their way into the planted areas.

Other than hand weeding, you can try the following:

- Cultivation—using a hoe, perform shallow soil disturbance when the weeds are small. This can bury them, reducing their access to light and exposing their roots, drying them up.
- Cutting off top growth—this reduces their food storage, but it may require a couple of rounds throughout the season—or even over a couple of years.
- String trimming—while quick for cutting top growth, string trimming makes it harder to be selective with which plants you are cutting.
- Mowing—this is a last resort if infestations are too well established for hand methods. Mow before weeds flower so they cannot set seed. This method may or may not work. Because you are not removing plants out of the ground, they have the potential to re-sprout.

Choose the best method for your access and always set up your roof for safe practices. Do not perform maintenance on rainy days, when there is a greater chance of slipping or on windy days, when materials will blow off the roof. Care should be taken on very hot days to avoid heat exhaustion. Try to weed first thing in the morning when temperatures can be a little cooler.

Plant Care

The same practices that you would follow for your at-grade gardens apply to your rooftop gardens. The level of maintenance you do will depend on your desired aesthetic. For example, grasses are best cut back in late winter or early spring, and flower heads can be cut back when blooming is complete. As green roofs are seen as refuges for birds, we tend to leave seed-producing flower heads throughout the winter so birds can eat the seeds. Also, if you have planted self-seeding plants for the purpose of rapid coverage, leave those flower heads alone until the spring.

You can divide plants as you would at-grade to fill in gaps, but you need to care for the divisions as you did the original plantings; you may even need to pamper them a bit more, as their root systems will be larger and they will require more water than the initial plugs. I usually bring extra growing media to maintenance visits. This way, if divisions or large weed infestation removals result in divots, they can be filled in.

Sedum can be cut back in the spring to remove spent flowers and encourage growth, but be careful how you do this. Using a line trimmer may seem efficient, but it can throw the cuttings off the roof; your neighbors may not be happy with where they land—and possibly take root. In addition, you'll be losing a valuable resource. You can use any cuttings to fill in gaps in your own plantings. Just be sure to give them their own establishment care. You should never walk on a sedum roof when frost is expected or after first frost because you can damage the leaf tissues of the plant.

If you have produced a large amount of debris after cutting back the plant growth, remove some of it, but not all. Too much dry debris can be a fire hazard or fly off the roof, but a thin layer left behind is beneficial in recycling nutrients back into the system.

Meadow Management

If you have purchased seeds from a seed supplier, they may have indicated that some species need to go through a cold spell in order to germinate. This is why meadows are best planted in the fall. The annual cover crop that is often supplied with the seeds will grow quickly to cover the soil to prevent erosion and reduce competition for weeds. It is important that the meadow is mowed in the spring *before* the annual cover crop has a chance to set seed. Mowing a meadow is beneficial because it encourages stronger root growth for the perennial wildflowers and it reduces the height of weeds and the annual cover crop, which brings light down to the soil. Mowing can be done with a lawn mower or a line trimmer.

Year 1

- A meadow requires mowing to a height of 4 to 6 inches. Mow throughout the first year whenever growth reaches 12 inches. Do this when the soil is dry to prevent soil compaction as you move around on the roof.
- In climates which receive snow, stop mowing in early fall. Leave some taller coverage to trap snow and insulate the soil. In climates with warmer winters, stop mowing when weeds stop producing flowers.

Year 2

- In the second year, mow in mid-spring. Mowing in mid-spring helps to set back nonnative cool-season weeds and grasses such as *Elymus repens* (quackgrass), *Poa annua* (annual bluegrass), and *Bromus* spp. (bromegrass). Mowing in mid-spring facilitates germination of dormant seed and enhances the growth of wildflowers without potentially increasing weeds. Mow right down as close as possible, and remove the cuttings. After this mid-spring mowing, you may not need to mow until the following spring, depending on how bad the weeds get.
- While fertilizing and watering is not typically recommended for naturalized meadow plantings, remember that on a roof, you may only have about 6–8 inches of artificial growing media—very different from your at-grade soil. The soil will heat up and lose water a lot faster than on the ground. For this reason, you may want to ensure you can get some water onto the roof in times of intense heat. Likewise, investigate the soil every year. If there does not seem to be any organic material, consider doing a light topdress or amendment. This is best done once you have good coverage with your meadow species; if you do it too soon, you will just encourage weedy competition.

Winterizing Your Roof

In terms of winterizing your roof, there is not much you need to do. Check to make sure the drains are clear of fallen leaves. If you have an irrigation system, you will need to blow out the lines and disconnect the system at the water source. Make sure none of your layering material or flashing has come loose and check for any damage from storms or animals. In the winter, do not use de-icing salts or compounds to melt the snow and ice. These can rust metal flashings, and the chemicals may not be compatible with the membrane; plus, they will make their way into snow melt off the roof.

Spring Start-Up

Connect your irrigation system if you have one, clear drains, and check for any winter damage or shifting of material at wall joints, edging, or drain outlets. Depending on your plant palette, now is the time to fertilize and cut back grasses and spent blooms.

Roof Checks

Every time you go on the roof to maintain the garden or open and close it for the season, take a look at the drainage outlets. Open lids to clear any material or creatures that may have made it in. If you have a grate system, chances are moss has begun to form along the face. While this is beautiful, it can slow the drainage of water from the roof. If this is causing problems (large weed infestations along the edge, water not draining), scrape it off. Throw it on the roof if you'd like. In addition to drainage outlets, make sure you do a thorough investigation of all other components: inspect the edging for shifts; check flashing on all penetrations and at walls; make sure there are no visible signs of wear, expansion, or contraction; and ensure all access points are safely in place without signs of damage.

As with growing on the ground, any garden requires a little bit of care to ensure long-term success. The best chance you have for a healthy green roof comes as a result of good establishment practices. Give your plants the best chance of survival by ensuring they receive adequate water, nutrition, and space. Addressing potential problems early, such as competition, poor soil conditions, or wear on flashing material, will reduce the resources and time required in the long run.

Fig. 10.3: *You should always visit your green roof, even if it's just to check your roof flashings and drains. This internal drain is not surrounded by clear stone; as a result, it filled with debris, preventing it from adequately draining.*
Photo Credit: Restoration Gardens, Inc.

Chapter 11
Food Production Roofs

Concerns over food security have prompted the growth of interest in rooftop food production. It is an obvious choice: roofs are typically large, underutilized surfaces with full sun exposure and low slopes. The biggest challenges are getting adequate access to the roof and a building's available weight load. You can plan a roof for food production, or, like one roof featured later in this chapter, you can convert an existing "passive" green roof into one that yields food. To create a food production roof, you should first understand the information given in Chapters 8 and 9 for creating a green roof. However, there are a few differences between a planted green roof and a food production roof.

- A Stronger Roof is Needed for Additional Loading Capacity
  - Unlike a 4-inch extensive roof, you will need to plan for a *minimum* of 6–8 inches if you want to grow a variety of produce. This provides greater water retention and nutrient availability, as well as a suitable depth. Depending on the roof design, some deeper substrates or even heavier pots can be placed around the perimeter of the roof or on point loads.
  - Live loads need to account for frequent maintenance personnel and potential visitors.
  - Maintenance of a food production roof requires small hand tools and/or buckets. Try to create a permanent safe spot for these on the roof to eliminate the need to repeatedly haul them up and down.

- Occupancy Permits May Be Required
  - Municipalities differ in regard to their permit requirements for installing a green roof. While passive green roofs with traditional plantings require only a few annual maintenance visits, food roofs require constant attention; production relies on a high frequency of visitations. Therefore, you must consult with your local municipality for any additional approvals required. Some jurisdictions may require changes to an occupancy permit and some may request zoning variances. Local building codes will dictate additional safety and fire compliance, such as permanent safety barriers and means of egress.

- Permanent and Clean Irrigation Is Needed
  - Permanent irrigation is required for these thirsty rooftop plants. For a small roof, a hose and nozzle are sufficient. For larger roofs, consider drip irrigation.
  - Harvesting rainwater is a sustainable option for watering plants. However, when you are watering edible plants, consider the following:
    - Avoid collecting water from treated wood shake roofs; these contain arsenic and elevated cadmium and copper. They can also leach phthalates.[1]
    - Avoid collecting from copper roofs or roofs with copper flashing, copper sulfate impregnated fabrics, or copper foil materials. The copper can easily leach into the collected rainwater.[2]

Fig. 11.1: *Safety railings and adequate loading is required for volunteers. As shown here, little hands are helpful on the Grame Rooftop Vegetable Garden in Montreal.*
PHOTO CREDIT: GRAME

- Collecting water from a green roof will recycle any nutrients that leach back into the system but may collect other leachates from roofing materials. While some studies have shown these will not affect your crops, you may want to do more research into your chosen products and look for materials that are food-grade.
- Place irrigation lines in the soil rather than irrigating plants directly. This will help filter any contaminants.
• Maintenance Pathways Are a Bigger Benefit
  - Instead of pathways only around the perimeter of the roof, vegetable rooftop gardens benefit from having pathways throughout. They allow you to maximize the growing spaces without compacting the soil as you move along. These extra pathways also reduce the spaces where weeds can grow.
• Environmental Exposures Are More Critical
  - Some varieties of leaf lettuce and herbs need some relief from the hot sun.
  - More delicate plants will need protection from the wind. Try wind-blocking with stronger plants or installing trellises.

- Soil Fertility Needs to Be Higher
  - Unlike the aggregate-based soils promoted for extensive green roofs, vegetable roofs will require more initial organics. In addition, you will need to provide a reliable organic amendment annually, perhaps even biennially. Both macro and micronutrients are vital for growing food. When I visited Eagle Street Rooftop Farm (New York) in 2011, they had their own composting system on the roof—complete with manure-producing chickens! At that point, Eagle Street's growing media contained a 40:60 percent ratio of organic to inorganic material[3]. In the Color Section, Photo 20, you can see a comparison of Ryerson Urban Farm's (Toronto) organic soil compared to some newly installed aggregate mix.
  - Consider the logistics when designing this type of roof. For example, how will you get materials onto the roof when they're needed? How can you situate compost piles so they sit over point loads?
  - Depending on the size of the garden and the intent for growing, some owners may want to do a thorough investigation of the soil nutrients by having it assessed by a lab. This will determine any deficiencies in the soil in terms of macro and micronutrients.
- Weeding Is a Must
  - Because weeds compete for nutrients, there is more of a need to weed a food roof than a traditional extensive or semi-intensive roof. Finding an efficient method will be key to keeping the workload light.

Suggested Crops

Some key lessons from those who grow on the roof include:

- To optimize available resources, favor plants with high yield that take up less space. Plant items like tomatoes, peppers, lettuce, and herbs—*not* squashes or watermelons.
- Plant items that can take advantage of height with trellises and supports—these will also provide much-needed shade structures on the roof, and/or they can be lined up to create windbreaks.
- If your access to the roof is not quick and effortless, do not plant your everyday kitchen items such as herbs and fresh lettuces on the roof.
- Choose dwarf varieties of crops if they are suitable for your climate.

Figure 11.2 lists popular crops, with media-depth recommendations.[4]

Figure 11.2: Recommended Crops with Recommended Media Depths

Crop	Depth (inches)
Lettuce	4–6
Spinach	4–6
Thyme	4–6
Oregano	4–6
Salvia	4–6
Chives	4–6
Onion	6
Beets	6
Hot Peppers	8
Carrots	8
Peas	8
Tomatoes	8

Growing in Rows

Food production roofs can be designed as built-in-place systems with rows or in containers, or both. For rows, build the roof as I have laid out previously, providing good drainage for the system. Leave yourself 12–18 inches of space to walk in or kneel on between rows; this will prevent damage to crops. These maintenance rows should consist of membrane, root barrier, drainage, and filter cloth—at a minimum. For an existing green roof that you are converting into a food production roof, if the roof's weight capacity allows, rake the existing growing media into rowed mounds. Growing in rows versus containers makes it easier to install efficient drip irrigation.

Growing in Containers

Containers can be wooden, metal, or plastic. If you use them, follow similar practices to those used for raised beds on the ground. Side walls contain the media, and there must be a way for water to drain without clogging. If you are making a lot of large custom wooden or metal frames, line the entire surface of the green roof (or at least the area where you are growing) with root barrier, drainage, and filter cloth, then place the bed frames on top. See Figure 11.3 for an example.

Fig. 11.3: *Easy-to-make lumber boxes can sit right on top of the green roof layers.*

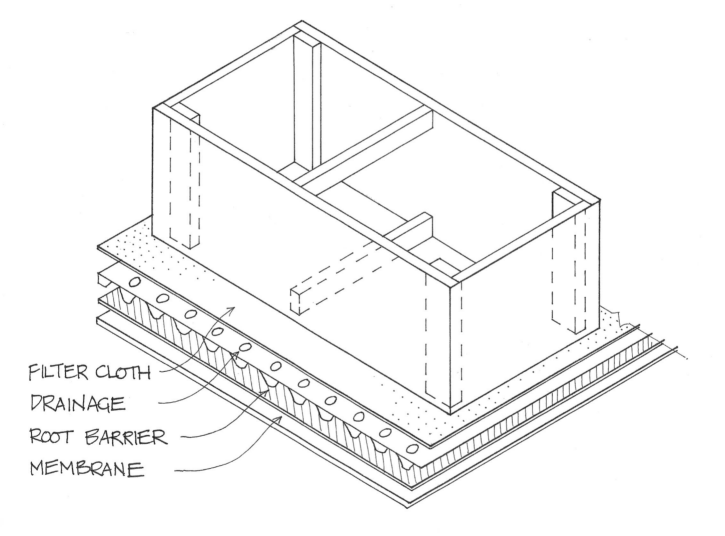

If you are growing food in smaller containers (i.e., pots or buckets) and leaving the majority of the membrane exposed, while you are producing food, your roof will not be classified as a green roof. This is because there will be large areas of the roof that do not provide the benefits of a traditional green roof, such as run-off reduction, heat reduction, and thermal properties. In addition, any membrane that is exposed will have to be replaced sooner than the areas that are covered. If you are planning on building a food-production roof and want to tap into any of your regional green roof incentives, check requirements for qualification.

Any perennial crop will need to be grown in a built-in-place system with an insulated bottom (such as the roof structure) or in insulated containers.

Container Materials

When building out of lumber, avoid pressure-treated wood. The chemicals that produce its rot-resistant properties should not be mixed with edible crops. For metal containers, choose something that is corrosion resistant, but keep in mind that metal containers will heat up and may dry your soil quickly. Galvanized containers will leach zinc, but not in quantities harmful to your food. Your local environmental authority may request that zinc-leaching products be avoided if you are near a sensitive body of water. Again, avoid the use of copper pots. However, if you already have copper pots and want to use them, consider lining them to prevent contact with your soil.

People grow a lot of food in plastic 5-gallon pails, and these could be used as accents in extensive roofs where the depth cannot be increased. You will have to find a way to allow these buckets (or other plastic containers) to drain. Some plastics are not UV stable, so check with the manufacturer for material properties. If you are concerned with products leaching chemicals, you may have to do a bit of further research or look for products that are designated as food grade.

You can find lots of self-watering systems in an ebook written by a Montreal group, *Alternatives: Guide to Setting up Your Own Edible Rooftop Garden.*[5] A self-watering system is ideal for thirsty fruiting vegetables; however, frequent, excessive watering can wash away nutrients and use up valuable resources.

If you will be growing in containers, you can use potting soil purchased from garden centers, but keep in mind that most contain peat moss, which is only very questionably-sustainable, and, once it dries out, it is very difficult to wet again. The type of potting soils you buy at the garden center balance drainage with water retention; they may include components such as vermiculite, perlite, and compost. Potting soil may also be labeled *potting mix, starting mix, growing mix,* or *transplanting mix.* Instead of buying bags of potting soil, you can make your own custom mix. As discussed previously (see Chapter 7), the priority is using lightweight materials, and you should avoid anything that already contains chemical fertilizers. Potting soil is not recommended for rows that are built-in-place on food roofs because it's too light and will blow away.

Garden soil is not recommended for containers because it tends to get compacted.

Food Roof Case Studies

Many food roofs or rooftop farms exist in North America. Most are located in major cities, including New York, Chicago, Montreal, Toronto, and Portland. Nonprofit organizations, urban farmers, and restauranteurs manage most rooftop farms as commercial ventures. Many of these growers offer tours and workshops so visitors can learn about the farming practices on the roof. Increasingly, condominium buildings are installing food roofs, giving tenants access to plots on shared accessible terraces. I had a hard time locating examples of food roofs on houses, but that does not mean they do not exist!

In this next section are three case studies of food production roofs. Although contractors built them, the food-production lessons and material components listed are helpful for the DIY builder.

Case Study: Grame Rooftop Vegetable Garden [6]

Location:
Montreal, QC

Zone:
USDA 5

Age of Roof:
Green roof installed in 2012, gardening started in 2013

Purpose:
Research, education, and environmental awareness

Weight Load Allowance for Green Roof:
15 psf for roof area that was not structurally reinforced and 45 psf for the reinforced area. 60-person capacity

Roof Size:
1,930 sq. ft with 1,000 sq. ft of growing space

System Type:
Built-in-place rows as well as containers

Layers:
- Root-resistant membrane
- "Eggshell" sheet drainage
- Expanded clay
- Geotextile for root control and filtering
- Growing media is a lightweight mix containing expanded clay, sourced from Canadian company Savaria

Depth of Growing Media and Crops:
- Pollinator garden, 6.5 inches: fruiting shrubs (gooseberry, cassis, chokeberry, haskap berries) and herbs
- Cultivated rows, 12 inches: tomatoes, lettuce, basil, garlic, peas, beans, zucchini (courgettes), and cucumbers; currently testing six varieties of strawberries
- Smart Pots, 15 gal: radishes, beets, cilantro, chives, strawberries

Unsuccessful Crops:
Rhubarb, melons, and pumpkins, due to lack of soil depth and available space

Yearly Soil Amendments:
Bagged compost and *Black Earth* (because soil volume decreases each year)

Irrigation System:
Beds are irrigated in hot weather with a drip system using potable water on a timer. Pots are hand watered.

What is one feature you cannot live without?
"Irrigation. Our crops suffered when the irrigation was down."

What is one wish you have for your food production roof?

"Wind protection, as we've experienced crop damage and dry soil."

Season:

Mid-May to end of October

Fig. 11.4: *In addition to growing food, the Grame Rooftop team is researching various species of strawberries in row planting for the green roof environment.*
Photo credit: Grame.

Case Study: Ryerson Urban Farm [7]

Managed by:
Arlene Throness

Location:
Toronto, ON

Zone:
USDA 5–6

Age of Roof:
Green roof installed in 2004, converted to a farm in 2013

Purpose:
Research, education, and local food production through a CSA

Weight Load Allowance for Green Roof:
52 psf green roof dead load allowance; 23 psf live load allowance

Roof Size:
9,000 sq. ft

Fig: 11.5: *What was once a passive green roof filled with daylilies was converted into a productive space by students at Ryerson University—accomplished, in large part, simply by raking the growing media into mounded rows.* Photo Credit: Leslie Doyle

System Type:

Built-in-place rows, with some self-watering containers

Layers:

- Membrane: 2-ply hot rubberized asphalt Hydrotech Monolithic Membrane with Hydroflex 30 separation sheet
- Root barrier: ZinCo WSF 40
- Insulation: two layers of 3-inch extruded polystyrene
- Drainage board: ZinCo Floradrain FD 60
- Additional drainage: crushed brick
- Filter cloth: ZinCo System Filter SF
- Original growing media: 150mm soil mix of organic and inorganic material made of crushed brick, shredded pine bark, blond peat, perlite, sand, and compost from vegetable matter
- Original plants: daylilies

Depth of Growing Media and Crops:

- Rows 10–12 inch deep: summer squash, cucumbers, salad greens, onions, garlic, eggplants, radishes, carrots, sweet peppers, turnips
- Self-watering containers: hot peppers

Arlene mentions that due to the more natural soil ecology of the rows, they are far more productive than the containers. The soil in the containers requires more amendments, and sometimes material has to be removed and replaced.

Yearly Soil Amendments:

For the first five years, 2 inches of compost was added to each row. This has resulted in an extremely high-organic mixture that is soft to the touch with ample aeration and water retention. A soil analysis indicated a deficiency of manganese. Manganese was added in a powdered form, and feather meal was added for nitrogen. Liquid seaweed and/or fish emulsions are applied as vegetables are fruiting.

Irrigation System:

Drip irrigation using potable water. Usually, only once a week, as organic material holds moisture well.

What is one feature you cannot live without?

"The fertile soil has led to our great success in producing an abundance of healthy produce."

What is one wish you have for your food production roof?

"Two wishes: additional safety fencing and a wash station for produce."

Growing Season:

April to November

Case Study: Noble Rot Restaurant [8,9]

Managed by Marc Bourcher-Colbert of Urban Agriculture Solutions LLC

Location:
Portland, Oregon

Zone:
USDA 8b–9a

Age of Roof:
Gardens were built in 2006

Purpose:
Fresh produce for the restaurant below

Weight Load Allowance for Green Roof:
20 psf deadload; built over steel trusses

Roof Size:
3,000 sq. ft

System Type:
Combination of planters and traditional built-in-place extensive green roofs.

A. Six large steel planter boxes double as railings (they were hoisted by crane).
B. Wooden planter boxes.
C. Two traditional green roofs, which add approximately 400 sq. ft of planted space.

Layers:

A. Steel planters with drains. Steel planters' legs are welded into the roof structure
B. Wooden planters sit on the of the modified bitumen membrane, lined with Firestone fPP-R Geomembrane, with a base of horticultural grade perlite covered in a geotextile. Drainage openings are at the sloped end.
C. The layers for the extensive green roof portion could not be confirmed. Suppliers unknown.

Growing Media:
Media in steel planters (A) and green roof area (C) were aggregate-based, with pumice. Wooden planters (B) have bagged potting soil containing coconut fiber and peat moss.

Depth of Growing Media:
- Steel planters: 24 inches
- Wooden planters: 6–8 inches
- Green roof area: 5–6 inches

Crops:
- Summer: beans, squash, eggplants, tomatoes, lettuces, herbs
- Winter: kale, brassica family, lettuces
- Perennial crops: thornless cactus, nasturtium, herbs

The extensive green roofs were started with Mediterranean herbs until the soil gained richness.

Marc notes that the tomatoes required lots of water, which was a challenge when there was a lot of transpiration on hot, windy days. He also found they sometimes did not taste as good as those grown on the ground.

Yearly Soil Amendments:
Aggregate-based soils were amended with liquid fertilizer for the first few years. Then Marc started a sheet mulching system. He throws his kitchen scraps onto one bed and covers it with felt fabric to contain it or pond liner to keep moisture in. The bed becomes full of life and workable compost within a year. He then plants in the bed containing the new fertile growing media and starts a new bed for mulching. The beds always have some residual scraps, which break down over time.

Marc uses basic vegan granular slow-release fertilizer (4-3-2, with micronutrients) and sometimes a powdered form of similar material. He also uses liquid fertilizers for quick results, when needed.

Three years ago, Marc started incorporating biochar into his spring fertilizer. Due to the limited supply, he would top dress to 50% of the square footage in cups (so a 20 sq. ft garden would receive 10 cups), then it would be mixed into the soil. This has been very successful, allowing him to finally grow the basil and arugula he wants; the herbs suffered before the biochar amendment.

Irrigation:
Water is supplied from the glacial well below the building. Marc waters everything by hand. He had to buffer the minerals, as they were quite high.

What is one feature you cannot live without?
"The learning I gained from taking the GRP training through Green Roofs for Healthy Cities. This gave me the confidence to grow safely on the 'urban mountaintop,' understanding the importance of maintaining the integrity of the roof membrane."

What is one wish you have for your food production roof?
"Safe access. Currently, it's via a steep ladder to a roof hatch. It can be quite the challenge when bringing up bagged material or bringing down harvests."

Fig. 11.6: *The Noble Rot Restaurant Rooftop consists of a mix of traditional extensive green roofs, raised beds on the roof deck, and perimeter high beds with trellises to support plants.*
PHOTO CREDIT: MARC BOURCHER-COLBERT

Summary

Building a green roof can be straightforward if the elemental building blocks are taken care of: a strong roof structure; a quality, watertight membrane; adequate drainage; and appropriate growing media and plants. The DIY builder should design a roof that is simple with easy access. The planting plan should include species with similar resource demands. And, above all, the roof should be safe.

While some European countries (Germany, for example) have green-roof-specific materials available in stores, in North America you will have to do some searching or get creative. Some of the large green roofing companies will require you to purchase their systems and may require a certified installer. Some green roof suppliers may have limited stock that does not work with your design, or they may require you to purchase rolls or sizes larger than you need. Look around your region's building stores for what is available before you finalize your design. Often, materials used in landscaping and construction can work—just be sure they fully satisfy your design requirements.

Building a green roof is a worthwhile investment, not just for your enjoyment and your building's performance, but for the community at large, including those creatures who will return to it year after year.

Appendix A:
A List of Common Standards and Guidelines

FLL, Landscape Development and Landscaping Research Society e.V.
Created the "Green Roof Guidelines, Guidelines for the Planning, Construction and Maintenance of Green Roofs" 2018 Edition. Available here: shop.fll.de/de/green-roof-guidelines-2018-download.html.

The American National Standards Institute (ANSI)
An organization that coordinates the development of U.S. voluntary national standards. These ANSI standards have been written by the Single Ply Roofing Industry

- ANSI/SPRI RP-14 2016 Wind Design Standard for Vegetative Roofing Systems, Approved September 9, 2016
- ANSI/SPRI VF-1 External Fire Design Standard for Vegetative Roofs, Approved May 11, 2017
- ANSI/SPRI VR-1 2018 Procedure for Investigating Resistance to Root or Rhizome Penetration on Vegetative Roofs, Approved June 11, 2018
- ANSI/SPRI ED-1 2019 Design Standard for Edge Systems Used with Low Slope Roofing Systems, Approved June 3, 2019

ASTM International
ASTM E2400 / E2400M - 19 Standard Guide for Selection, Installation, and Maintenance of Plants for Vegetative (Green) Roof Systems
astm.org/Standards/E2400.htm

Factory Mutual Insurance Company
FM Global Property Loss Prevention Data Sheet 1-35 "Vegetative Roof Systems," February 2020.
fmglobal.com/research-and-resources/fm-global-data-sheets

Green Roofs for Healthy Cities (GRHC)
The North American industry association for the green roof (and green wall) sectors. GHRC creates awareness of the industry and the benefits of green roofs, works with cities to create local policies, co-author standard documents, and provides courses for professional accreditation (Green Roof Professionals, GRP) and professional development. greenroofs.org

Appendix B:
North American Cities with Green Roof Programs

This Appendix lists North American Cities with Green Roof Programs in place as of early 2021 per Green Roof for Healthy Cities *Green Roof and Wall Policy in North America*. Those that have mandatory requirements or those that offer incentive programs divide the list. An asterisk indicates programs that may be applicable to the DIY builder.

Mandatory Green Roof Requirements

Chicago, Illinois: Sustainable Development Policy, 2017

Denver, Colorado: Green Building Ordinance, 2017

Devens, Massachusetts: Policy for Construction of Vegetated Roofs, 2012*; Industrial Performance Standards and General Regulations: Greenhouse Gas Mitigation (974CMR 4.11 2 (c)), 2012; Viewshed Overlay District Vegetated Rooftops and Vegetated Walls Requirement (974CMR 3.04 (8) (i)5), 2013*;

Fife, Washington: Green Factor, 2009*

New York City, New York: Green Roofs for New Construction, 2019; Green Roofs on Smaller Buildings, 2019*

Portland, Oregon: EcoRoof Requirement, 2018

Saint Laurent, Quebec: Regluement sur le zonage no RCA08-08-001, 2016*

San Francisco, California: Better Roofs Ordinance, 2017*; Central SOMA Plan Living Roof and Solar Requirement, 2019*

Seattle, Washington: Green Factor, 2007*; RainWise Program, 2009

Toronto, Ontario: Green Roof Bylaw, 2009*

Washington, DC: Green Area Ratio, 2017*

Incentive Programs

Austin, Texas: Downtown Density Bonus Program, 2014

Chicago, Illinois: Green Permit Program, 2014*; Floor Area Bonus, 2015

Guelph, Ontario: Stormwater Credit and Rebate Program, 2017

Hoboken, New Jersey: Green Infrastructure Bonus Standards for Impervious Coverage, 2018*

Marion County, Indianapolis: Stormwater Credit, 2016*

Milwaukee, Wisconsin: Green Infrastructure Partnership, 2019*

Minneapolis, Minnesota: Stormwater Utility Fee Credit, 2017*

Montgomery County, Maryland: Rainscapes Rewards Rebate Program, 2018*

Nashville, Tennessee: Green Roof Rebate Program, 2016*

New York State: Green Roof Property Tax Abatement Program, 2019*

New York City, New York: Green Infrastructure Grant Program, 2011*; Green Roof Information, 2019*

Northeast Ohio Regional Sewer District: Impervious Area Reduction Credit, 2016*

Onondaga County, New York: Green Improvement Fund, 2018

Palo Alto, California: Stormwater Measures Rebate Program, 2017*

Philadelphia, Pennsylvania: Green Roof Tax Credit, 2016; Density Bonus, 2015; Stormwater Grants, 2018; Stormwater Credits Program and Incentives

Portland, Maine: Stormwater Credit, 2015*

Prince George County, Maryland: Rain Check Rebate Program, 2013*

Seattle, Washington: Land Use Code

Toronto, Ontario: Eco-Roof Incentive Program, 2009*

Washington, DC: Stormwater Retention Credit Training Program, 2013*; Riversmart Rewards Program, 2013*; Clean Rivers Impervious Area Charge Incentive Program*; Riversmart Green Roof Rebate Program, 2016*

Kitchener, Ontario: Stormwater Credit Policy, 2012*

Green Roof and Wall Policy in North America *Regulations, Incentives, and Best Practices (2019); Green Roofs for Healthy Cities.*

Credit: This report was prepared by Maya Stern, Program Coordinator, Green Roofs for Healthy Cities (GRHC); Steven W. Peck, GRP, Honorary ASLA, Founder and President, GRHC; and Jeff Joslin, Director, Current Planning, City of San Francisco.

Appendix C:
Relevant Climate Links

2000 Canadian Hardiness Zone Map via the Government of Canada website planthardiness.gc.ca

2012 United States Hardiness Zone Map via the United States Department of Agriculture website https://planthardiness.ars.usda.gov/PHZMWeb/

US Precipitation Map
Source: AHPS Precipitation Analysis, National Weather Service (weather.gov)

Canadian Precipitation Map
Source: Accumulated Precipitation, Agriculture and Agri-Food Canada (agr.gc.ca)

Endnotes

Chapter 3

1. Mark T. Simmons, "Climates and Microclimates: Challenges for Extensive Green Roof Design in Hot Climates," in *Green Roof Ecosystems,* ed. by Richard K. Sutton (Switzerland: Springer International, 2015) 68, 71, 73–76.
2. Toronto Municipal Code Chapter 492, (Toronto: 2017) 492–2 and 492–4.
3. Charles D. Miller, "Green Roofs: A New American Building System," *Building Safety Journal,* July 22, 2019, iccsafe.org/building-safety-journal/bsj-technical/green-roofs-a-new-american-building-system
4. Factory Mutual Insurance Company, "Vegetative Roof Systems," FM Global Property Loss Prevention Data Sheets, February 2020, 8.
5. "Sedum FeatherMats (SMUL6)" Sedum Master, accessed: September 1, 2020, sedummaster.com/Sedum-Feather-Mats.htm
6. "Systems with Drainage Mats," Conservation Technology, accessed September 1, 2020, conservationtechnology.com/greenroof_systems_mats.html
7. "Terrafix 360R—Geotextile," Terrafix Geosynthetics Inc. terrafixgeo.com/wp-content/uploads/360R-_-Light-Weight-Nonwoven-Geotextile.pdf
8. "System and Components for Roof Landscapes: Product List 2016/2017," ZinCo GmbH, 9.
9. "Styrofoam™ Brand Roofmate™ Extruded Polystyrene Foam Insulation," DuPont Product Information Sheet.
10. "System and Components for Roof Landscapes: Product List 2016/2017," ZinCo GmbH, 4.
11. "Liberty™ SBS Self-Adhering Roo ng System," GAF, gaf.ca/roo ng/residential/products/ low_slope_membrane/liberty
12. Forschungsgesellschaft Landschaftsentwicklung Landschaftsbau (FLL) e.V, "Green Roof Guidelines: Guidelines for the Planning, Construction and Maintenance of Green Roofs," 2018 Edition. Landscape Development and Landscaping Research Society e.V., 46–50.
13. ZinCo GmbH, " Planning Guide: Systems for Pitched Green Roofs," Brochure, 5.

Chapter 4

1. "Extension Ladder Chart," Sunset Ladder, last Modified October 16, 2017, sunsetladder.com/blog/extension-ladder-chart
2. Section 26.1, Ontario Regulation 213/91, Construction Projects, Occupational Health and Safety Act, R.S.O. 1990, c 0.1. ontario.ca/laws/regulation/910213

Chapter 5

1. "Wind Design Standard for Vegetative Roofing Systems," ANSI/SPRI, 2016, 7, 44.

Chapter 6

1. Mathis Natvik, "Ecosystems as Models for Plant Selection on Extensive Green Roofs in Southern Ontario" (Thesis, University of Guelph, 2012), 59.

2. Scott Torrance et al., "City of Toronto Guidelines for Biodiverse Green Roofs" (Toronto City Planning, 2013), 33.
3. Factory Mutual Insurance Company, "Vegetative Roof Systems," FM Global Property Loss Prevention Data Sheets, February 2020, 9.
4. Scott Torrance et al., "City of Toronto Guidelines for Biodiverse Green Roofs" (Toronto City Planning, 2013), 13–15.
5. Jeremy T. Lundholm, "Green Roof Plant Species Diversity Improves Ecosystem Multifunctionality," *Journal of Applied Ecology*, 52 (2015), 726.

Chapter 7

1. Takanori Kuronuma et al., "CO_2 Payoff of Extensive Green Roofs with Different Vegetation Species," *Sustainability* 10 (2018): 2256.
2. Fabricio Bianchini and Kasun Hewage, "How 'Green' Are the Green Roofs? Lifecycle Analysis of Green Roof Materials," *Building and Environment* 48 (2012) 57–65 doi:10.1016/j.buildenv.2011.08.019
3. Graham, 2007 as cited by Edmund C. Snodgrass and Linda McIntyre, *The Green Roof Manual: A Professional Guide to Design, Installation, and Maintenance*. (Portland: Timber Press, Inc., 2010) 89.
4. "Waterproofing Considerations for Green Roofs," Features, Construction Canada, March 1, 2012, constructioncanada.net/waterproofing-considerations-for-green-roofs/2/
5. Kelly Luckett, "Green Roof Construction and Maintenance" (New York: McGraw-Hill's GreenSource, 2009), 40.
6. Factory Mutual Insurance Company, "Vegetative Roof Systems," FM Global Property Loss Prevention Data Sheets, February 2020, 11.
7. Forschungsgesellschaft Landschaftsentwicklung Landschaftsbau (FLL) e.V, "Green Roof Guidelines: Guidelines for the Planning, Construction and Maintenance of Green Roofs," 2018 Edition. Landscape Development and Landscaping Research Society e.V., 69.
8. Cuong T.N. Cao et al., "Biochar makes green roof substrates lighter and improves water supply to plants" *Ecological Engineering* 71 (2014): 368–374.
9. Susan Parent, "Criteria for Growing Media for Roof Top Gardening," Training Centre PROMIX, September 29, 2020, pthorticulture.com/en/training-center/criteria-for-growing-media-for-roof-top-gardening
10. Friedrich, C.R., in "Principles for selecting the proper components for a green roof growing media," 2005 as cited by Nigel Dunnett and Noel Kingsbury, *Planting Green Roofs and Living Walls, Revised and Updated Edition*. Portland: Timber Press, 2008.
11. Canadian Sphagnum Peat Moss Association (CSPMA), "Research" in Responsible Production. Accessed September 29, 2020. peatmoss.com/responsible-production/research-the-key-to-integrated-sustainable-management
12. Factory Mutual Insurance Company, "Vegetative Roof Systems," FM Global Property Loss Prevention Data Sheets, February 2020, 8.
13. Forschungsgesellschaft Landschaftsentwicklung Landschaftsbau (FLL) e.V, "Green Roof Guidelines: Guidelines for the Planning,

Construction and Maintenance of Green Roofs," 2018 Edition. Landscape Development and Landscaping Research Society e.V., 46–50.
14. "Green Home Remodel: Healthy Homes for a Healthy Environment," Seattle Public Utilities, Sustainable Building Program. Accessed: October 4, 2020.
15. Factory Mutual Insurance Company, "Vegetative Roof Systems," FM Global Property Loss Prevention Data Sheets, February 2020, 13.
16. Toronto Municipal Code Chapter 492, (Toronto: 2017) 492–11.

Chapter 11

1. "Roofing Materials Assessment: Investigation of Toxic Chemicals in Roof Runoff," Nancy L. Winters and Kyle Graunke, (Washington: Washington State Department of Ecology, February 2014), 9.
2. "Green Home Remodel: Healthy Homes for a Healthy Environment," Seattle Public Utilities, Sustainable Building Program. Accessed October 4, 2020.
3. Annie Novak, *The Rooftop Growing Guide: How To Transform Your Roof into a Vegetable Garden or Farm.* (New York: Ten Speed Press) 2016, 89.
4. Germain, Amelia, et al. *Guide to Setting up Your Own Edible Rooftop Garden* (Montreal: Alternatives and the Rooftop Garden Project) 2008, 45.
5. Germain et al., *Edible Rooftop Garden*, 34.
6. Femke Bergsma of Grame, email and phone communication, July 21, 2020.
7. Arlene Throness, Manager, Urban Farm, University Business Services, Ryerson University, email and phone communication, July 21, 2020.
8. Marc Boucher-Colbert, Urban Farmer and Garden Specialist, email and phone communication, July 24, 2020.
9. Benjamin Engelhard, "Rooftop to Tabletop: Repurposing Urban Roofs for Food Production" (Thesis, University of Washington, 2010) 46–50.

Bibliography

Agra, Har'el, et al., "Measuring the Effect of Plant-Community Composition on Carbon Fixation on Green Roofs," *Urban Forestry & Urban Greening* 24 (May 2017): 1–4. doi.org/10.1016/j.ufug.2017.03.003

ANSI/SPRI Standard RP-14, "Wind Design Standard for Vegetative Roofing Systems," September 9, 2016.

Appleby-Jones, S., J. Lundholm, and A. Heim. "Kelp Extract (Ascophyllumnodosum) Can Improve Health and Drought Tolerance of Green Roof Plants," *Journal of Living Architecture* 4(1)(2017): 1–13.

Bates, Adam J., et al., "Effects of Varying Organic Matter Content on the Development of Green Roof Vegetation: A Six Year Experiment," *Ecological Engineering*, 82 (2015): 301–310. doi.org/10.1016/j.ecoleng.2015.04.102

Beck, Deborah A., Gwynn R. Johnson, and Graig A. Spolek, "Amending Greenroof Soil with Biochar to Affect Runoff Water Quantity and Quality," *Environmental Pollution* 159 (2011): 2111–2118. doi:10.1016/j.envpol.2011.01.022

Belair Direct. "Designing and Insuring a Green Roof," Belair Direct blog. Accessed September 2020. blog.belairdirect.com/designing-insuring-green-roof/

Bianchini, Fabricio and Kasun Hewage. "How 'Green' Are the Green Roofs? Lifecycle Analysis of Green Roof Materials," *Building and Environment* 48 (2012) 57–65. doi:10.1016/j.buildenv.2011.08.019

Bozorg Chenani, Sanaz, Susanna Lehvavirta, Tarja Hakkinen. "Life Cycle Assessment of Layers of Green Roofs," *Journal of Cleaner Production* 90 (2015): 153–162. dx.doi.org/10.1016/j.jclepro.2014.11.070

Canadian Roofing Contractors Association. "Ice on Roofs," Bulletin, Canadian Roofing Contractors Association. January 1998.

Canadian Sphagnum Peat Moss Association (CSPMA). "Research" in Responsible Production. Accessed September 29, 2020. peatmoss.com/responsible-production/research-the-key-to-integrated-sustainable-management/

Cao, Cuong T.N., et al., "Biochar Makes Green Roof Substrates Lighter and Improves Water Supply to Plants," *Ecological Engineering* 71 (2014): 368–374.

Chen, Haoming, et al., "Effects of Biochar and Sludge on Carbon Storage of Urban Green Roofs," *Forests* 9(7) (2018) 413.

Circular Ecology, "Embodied Carbon: The ICE Database." Inventory of Carbon and Energy (ICE) 3.0, University of Bath, Nov 10, 2019. circularecology.com/embodied-carbon-footprint-database.html

Credit Valley Conservation, "Native Prairie and Meadow Gardens and Landscapes for Homes, Businesses and Institutions," Credit Valley Conservation. Accessed September 29, 2020. cvc.ca/wp-content/uploads/2013/05/12-205-prairiemeadow-booklet-web.pdf

Conservation Technology, "Systems with Drainage Mats." Accessed September 1, 2020.

Construction Canada, Allen Lyte, et al., "Critical Considerations for Replacing a Flat Roof," Features, March 9, 2013.

Canada Roofing Contractors Association. "Roof Re-Covering: An Alternative?" Bulletin, 48, June 1999.

Construction Canada. "Waterproofing Considerations for Green Roofs," Features, March 1, 2012.

Dunnett, Nigel, et al., *Small Green Roofs: Low-Tech Options for Greener Living.* Portland: Timber Press, Inc. 2011.

Dunnett, Nigel and Noel Kingsbury. *Planting Green Roofs and Living Walls, Revised and Updated Edition.* Portland: Timber Press, 2008.

DuPont. "Styrofoam™ Brand Roofmate™ Extruded Polystyrene Foam Insulation," Product Information Sheet.

Ecohome. "Rigid Insulation Panels: Which Ones to Use for Different Applications." Accessed September 2020.

Emilsson, Tobias. "Vegetation Development on Extensive Vegetated Green Roofs: Influence of Substrate Composition, Establishment Method and Species Mix," *Ecological Engineering* 33 (2008) 265–277.

Factory Mutual Insurance Company. "Vegetative Roof Systems," FM Global Property Loss Prevention Data Sheets, February 2020.

Fletcher, Michael. *Moss Grower's Handbook: An Illustrated Beginner's Guide to Finding, Naming and Growing over 100 Common British Species.* Berkshire: Seven Ty Press, 1991.

Forschungsgesellschaft Landschaftsentwicklung Landschaftsbau (FLL) e.V. "Green Roof Guidelines: Guidelines for the Planning, Construction and Maintenance of Green Roofs," 2018 edition. Landscape Development and Landscaping Research Society e.V.

Fulthorpe, Roberta, et al., "The Green Roof Microbiome: Improving Plant Survival for Ecosystem Service Delivery," *Frontiers in Ecology and Evolution* 6:5 (2018). doi:10.3389/fevo.2018.00005

GAF. "Liberty™ SBS Self-Adhering Roofing System." gaf.ca/roofing/residential/products/low_slope_membrane/liberty

Germain, Amelie, et al., "Guide to Setting up Your Own Edible Rooftop Garden," Montreal: Alternatives and the Rooftop Garden Project, 2008.

Gedge, Dusty and John Little. *The DIY Guide To Green and Living Roofs,* ebook, 2008.

Glime, Janice M., *Bryophyte Ecology.* Volume 1. Physiological Ecology. ebook sponsored by Michigan Technological University and the International Association of Bryologists, 2007. Accessed April 2020. bryoecol.mtu.edu

Green Roofs for Healthy Cities. *Green Roof Waterproofing and Drainage* 301, participants manual, 2007.

Hill, Jenny, Jennifer Drake, and Brent Sleep. "Comparisons of Extensive Green Roof Media in Southern Ontario," *Ecological Engineering* 94 (2016): 418–426. dx.doi.org/10.1016/j.ecoleng.2016.05.045

IKO. "Guide to Drip Edges for Shingle Roofs: Is a Drip Edge Necessary?" Learn About Roofing. Accessed September 2020. iko.com/na/pro/building-professional-tools/learn-about-roofing/guide-to-drip-edges-for-shingle-roofs

Ireland R., R. Cain. *Checklist of Ontario Mosses.* National Museums of Canada, Publications in Botany No. 5. ON: Ottawa, 1975.

Ireland R., L. Ley. *Atlas of Ontario Mosses.* Ottawa: Canadian Museum of Nature, 1992

John, Jesse, Gavin Kernaghan, and Jeremy Lundholm. "The Potential for Mycorrhizae to Improve Green Roof Function," *Urban Ecosystem* 20 (2017): 113–127. doi.org/10.1007/s11252-016-0573-x

Kuronuma, Takanori, et al., "CO_2 Payoff of Extensive Green Roofs with Different Vegetation Species," *Sustainability* 10 (2018): 2256. doi:10.3390/su10072256

Lubell, Jessica D., Kris J. Barker, and George C. Elliott. "Comparison of Organic and Synthetic Fertilizers for Sedum Green Roof Maintenance," *Journal of Environmental Horticulture* 31 (4) (2013): 227–233.

Luckett, Kelly. *Green Roof Construction and Maintenance.* New York: McGraw-Hill's GreenSource. 2009

Lundholm, Jeremy T. "Green Roof Plant Species Diversity Improves Ecosystem Multifunctionality," *Journal of Applied Ecology* 52 (2015), 726–734. doi: 10.1111/1365-2664.12425

Miller, Charles D. "Green Roofs: A New American Building System," *Building Safety Journal,* July 22, 2019. iccsafe.org

Missouri Botanical Gardens. "Plant Finder" Accessed 2020. missouribotanicalgarden.org/plantfinder/plantfindersearch.aspx

Molineux, Chloe J., Alan C. Gange, and Darryl J. Newport. "Using Soil Microbial Inoculations to Enhance Substrate Performance on Extensive Green Roofs," *Science of the Total Environment* 580 (2017); 846–856. dx.doi.org/10.1016/j.scitotenv.2016.12.031

Natvik, Mathis. "Ecosystems as Models for Plant Selection on Extensive Green Roofs in Southern Ontario" Thesis, University of Guelph, 2012.

Newcomb, L. *Newcomb's Wildflower Guide.* New York: Little Brown, 1977.

Novak, Annie. *The Rooftop Growing Guide: How to Transform Your Roof into a Vegetable Garden or Farm.* New York: Ten Speed Press, 2016.

Occupational Health and Safety Act, R.S.O. 1990, c 0.1., *Ontario Regulation 213/91, Construction Projects, Section 26.1.* ontario.ca/laws/regulation/910213

Olszewski, Michael W. and Sasha W. Elsenman."Influence of Biochar Amendment on Herb Growth in a Green Roof Substrate," *Horticulture, Environment, and Biotechnology* 58(4) (2017): 406–413. doi.org/10.1007/s13580-017-0180-7

Olszewski, Michael W., Marion H. Holmes, and Courtney A. Young. "Assessment of Physical Properties and Stonecrop Growth in Green Roof Substrates Amended with Compost and Hydrogel," *Technology and Products Reports, Hort Technology,* April 20 (2010): 438–444.

Onodono, S., J.J. Martinez-Sanchez, and J.L. Moreno. "The Inorganic Component of Green Roof Substrates Impacts the Growth of Mediterranean Plant Species as Well as the C and N Sequestration Potential,"

Ecological Indicators 61 (2016): 739–752. dx.doi.org/10.1016/j.ecolind.2015.10.025

Patton, Dennis. "Defining Sun Requirements for Plants: Shedding Light on Sun/Shade Conditions," K-State Research and Extension. johnson.k-state.edu/lawn-garden/agent-articles/miscellaneous/defining-sun-requirements-for-plants.html

Parent, Susan. "Criteria for Growing Media for Roof Top Gardening," Training Centre PROMIX, September 29, 2020. pthorticulture.com

Quinty, Francois and Line Rochefort. *Peatland Restoration Guide, Second Edition.* Quebec: Canadian Sphagnum Peat Moss Association and New Brunswick Department of Natural Resources and Energy, 2003.

Schofield W.B. *Introduction to Bryology.* New York: Macmillan, 1985.

Seattle Public Utilities. "Roofing," Green Home Remodel: Healthy Homes for a Healthy Environment, Seattle Public Utilities' Sustainable Building Program. Accessed October 4, 2020.

SedumMaster. *2018 Plant Catalogue,* online. Accessed September 29, 2020. sedummaster.com

Simmons, Mark T. "Climates and Microclimates: Challenges for Extensive Green Roof Design in Hot Climates," in *Green Roof Ecosystems,* edited by Richard K. Sutton, 63–80, Switzerland: Springer International, 2015.

Snell, Clarke and Tim Callahan. *Building Green: A Complete How-To Guide to Alternative Building Methods.* New York: Lark Books, 2005, 178–194, 484–511.

Snodgrass, Edmund C. and Linda McIntyre. *The Green Roof Manual: A Professional Guide to Design, Installation, and Maintenance.* Portland: Timber Press, Inc., 2010.

Snodgrass, Edmund C. and Lucie L. Snodgrass. *Green Roof Plants: A Resource and Planting Guide.* Portland: Timber Press, 2006.

Stiffler, Lisa. "A Green Light for Using Rain Barrel Water on Garden Edibles," Sightline Institute, January 7, 2015. sightline.org

Studler, Susan. "Extensive Green Roofs and Mosses: Reflections from a Pilot Study in Terra Alta, West Virginia," *Evansia*, 26(2) (2009): 52–63.

Sunset Ladder. "Extension Ladder Chart," Blog. Last Modified October 16, 2017. sunsetladder.com

Sutton, Richard K., ed. *Green Roof Ecosystems,* Ecological Studies 223, Switzerland: Springer International, 2015. doi.org/10.1007/978-3-319-14983-7

Terrafix Geosynthetics Inc. "Terrafix 360R – Geotextile." terrafixgeo.com/wp-content/uploads/360R-_-Light-Weight-Nonwoven-Geotextile.pdf

Toronto Municipal Code. Chapter 492, "Green Roofs." Toronto: November 9, 2017. Accessed April 2020. toronto.ca/legdocs/municode/1184_492.pdf

Toronto and Regional Conservation Authority. "Native Plant Suggestions," Green Roof Pilot Program, 2007, City of Toronto.

Torrance, Scott, Brad Bass, Scott MacIvor, and Terry McGlade. "City of Toronto Guidelines for Biodiverse Green Roofs," Zoning Bylaw and Environmental Planning, Toronto City Planning, 2013.

Whittinghilla, Leigh, J. D. Bradley Rowe, Robert Schutzki, and Bert M. Cregg. "Quantifying Carbon Sequestration of Various Green Roof and Ornamental Landscape Systems," *Landscape and Urban Planning* 123 (2014): 41–48. dx.doi.org/10.1016/j.landurbplan.2013.11.015

Wildflower Farm. "Post-Seeding Management," How to Grow a Meadow, Wildflower Farm, Accessed September 2020. wildflowerfarm.com/index.php?route=information/information&information_id=932

Winters, Nancy L. and Kyle Graunke. "Roofing Materials Assessment: Investigation of Toxic Chemicals in Roof Runoff," Washington: Washington

State Department of Ecology, February 2014. Accessed July 29, 2020. resources.rainharvestingsupplies.com.

Young, Thomas M., Duncan D. Cameron, and Gareth K. Phoenix. "Increasing Green Roof Plant Drought Tolerance through Substrate Modification and the Use of Water Retention Gels," *Urban Water Journal*, 14:6 (2017): 551–560. doi.org/10.1080/1573062X.2015.1036761

ZinCo GmbH. "Planning Guide: Systems for Pitched Green Roofs." Company brochure.

ZinCo GmbH. "System and Components for Roof Landscapes: Product List 2016/2017." Company brochure.

Index

Page numbers in *italics* indicate diagrams and tables.

A
Abbey Gardens, Haliburton, Ontario, 83
access, 19–20, 27, 29–30
aggregate (drainage), 74, 76
aggregate (growing media), 79, 80, *81*
alpine plant species, 43
aluminum, 93–94
alvars, 43
annual plants, 45

B
biochar, 79–80, 94
biodiversity, 58–59
building codes, 14–15, 19, 123
Building Green (Snell, Callahan), 103, *104*

C
Callahan, Tim, 103
cant strips, 10, *10*, 112
carbon footprint, 63
carbon sinks, 63
Centre Hastings, Ontario, 97
climate, 38–39
coconut coir, 81
cold roof assemblies, 86
containers, 126–127, *126*
contractors, 33
conventional roof assemblies, 65, 86
copper, 94
copper sulfate impregnated fabrics, 71–72
crickets, 10, *11*
cuttings, 55, *57–58*

D
dead load, 15
design
    food production roofs, 123–127
    goals, *14*
    for maintenance, 60
    multi-purpose materials, 94
    plant biodiversity, 58–59, *59*
    rooftop measurements, 35–37, *35*, *36*
    site factors, 37–41
dimple board, 74–75, 94, 102, 110
drainage layers
    about, 7, 8, *8*, *64*
    filter cloth with, 77
    installation, 102–103
    multi-purpose materials, 94
    properties comparison, *76*
    role, 72–73
    slope considerations, 25
    specifications, 73–76
drainage outlets, 8, *8*
drainage sheet systems, 76
drain boxes, 73–74, *73*
drain edges, 101, 103–104, *104*
drip edges, 11, *12*, 93–94

E
Eagle Street Rooftop Farm, New York, 125
eaves, 12, *12*
eavestroughs, 12, *12*
EcoSuperior Garden Shed, Thunder Bay, Ontario, 84
edging, 92
emergency overflows, 10, *11*, 74
EPDM (ethylene propylene diene monomer) membrane, 67, 68, *70*, 85, 99–100
erosion control, 87–88, 94
extensive systems
    about, 2–3
    characteristics of, *3*
    weight loads, *16*, *17*
extruded polystyrene (XPS), 86–87

F

fertilization, 116–117, 125
filter cloth
    about, 7, 8, *64*
    drainage layer with, 74
    installation, 103, 110
    role, 76
    specifications, 76–77
flashing, 92–93, *93*
flat roofs, 16
Fleming College, 97
flood test, 65–66
flowing-water test, 66
fluid-applied membranes, 68, *70*
FM Global, 16
food production roofs
    case studies, 128–133
    crops, 125
    growing design, 126–127, *126*
    requirements, 123–125
functions, 21

G

gable roofs, 9
geotextile. *See* filter cloth
goals, *14*
Grame Rooftop Vegetable Garden, Montreal, Quebec, 128–129
green roofs
    benefit of, 1, 13
    carbon footprint, 63
    categories, 2–3
    functions, 21
    growing environment, 43–44
    new vs retrofit, 3–4
green roof systems
    elements of, 1–2
    layers, 7–9
groundcover plants, 46
growing media
    about, 7, 9, *64*
    components, 77–78, 80, *81*
    environmental concerns, 80–82
    fertilization and, 116–117
    food production roofs, 125
    installation, 103, 111, 113
    load capacity and, 21
    products, 79–80
    role, 77
    slope slippage, 88–90
    specifications, 85
    transport to roof, 30, 83, 85
    using soil, 82–83
    weights, 16
gutters, 12, *12*

H

home insurance, 15
hydrogels, 87, 88, 94

I

inside corners, 113, *113*
installation
    material transport to roof, 30, 83, 85
    new build case study, 97–105, *98*
    retrofit build case study, 107–113, *108*
    slope considerations, 26
    strategies, 22
insulation
    about, 9
    installation, 110
    multi-purpose, 94
    role, 85–86
    specifications, 86–87
insurance, 15
intensive systems, 2–3, *3*, 16
    See also food production roofs
invasive plant species, 53
irrigation
    about, 9
    food production roofs, 123–124
    maintenance, 116, 121
    methods for, 91
    with modular units, 95

slope considerations, 25–26

L
ladders, 27, *28*
layers
　installation, 112–113
　multi-purpose materials, 94
　optional, 9, 63
　quantity required, 95
　required, 7–9, 63
　slippage, 88–90
leak detection tests, 65–66
liquid-applied membranes, 68, *70*
live load, 15

M
maintenance
　access, 30
　design considerations, 60
　fertilization, 116–117
　frequency, 115
　irrigation, 116
　plants, 119–120
　roof inspections, 121
　slope considerations, 26
　weed control, 117–119
　winterizing, 121
manufactured aggregates, 80, *81*
materials
　amount required, 95
　finding supplies, 134
　multi-purpose, 94
　transport to roof, 30, 83, 85
mats (plants), 55, 56, *57–58*, 88
meadow plantings, 43, 120
mechanical systems, 37
membrane protection layer, 9, 85
mobile elevated work platforms, 30
modified bitumen membranes, 67, 68, *70*
modular units (plants), 55, 56, *57–58*, 95
moss roofs, 53–55
municipal regulations, 14–15, 123

N
native plants, 44, 46
naturalized green roofs, 60
Natvik Design, Inc., 97
new build case study, 97–105, *98*
Noble Rot Restaurant, Portland, Oregon, 132–133

O
occupancy, 19–20, 123
Ontario, codes, 20, 31, 119

P
parapets
　about, 9–10, *10, 11*
　emergency overflow, 74
　filter cloth on, 77
　flashing, 92, *93*
　height, 112
　inside corner membrane installation, *113*
　installation, 109
pathways, 94, 124
pea gravel, 94
penetrations, in roof, 37, 92–93, *93*, 112
perennial plants, 45
pesticides, 119
Pickering, Ontario, farm, 84
pig ear bend, *113*
plant hardiness zones, 38–39
plants
　about, 7, 9, 64
　biodiversity, 58–59
　characteristics required, 44–45
　design considerations, 58
　establishment period, 61
　fertilization, 116–117
　food production crops, 125
　growing environment for, 43–44
　installation, 103, 111
　load capacity and, 21
　maintenance, 119–120

moss, 53–55
planting methods, 55–56, *57–58*, 95
planting time, 60–61
recommendations, 46, *47–52*
slope considerations, 26
unsuitable species, 53
weed control, 117–119
weight loads, 58
plugs (plants), 55–56, *57–58*, 103, 111
point loads, 18–19
porch roofs, 65, 86, 97–105, *98, 104*
potting soil, 127
prairie plant species, 43
precipitation, 38–39, 87
protected membrane roof assemblies (PMR), 65, 86, *86*, 107
PVC single-ply membranes, 68

R

rain, 38–39, 87
rake edges, 11, *12*
replacement/repair strategies, 22–23, 30
Restoration Gardens, Inc., 107
retrofit build case study, 107–113, *108*
roadside plant species, 43–44
roof deck, 9, 98–99, 109
roof hatch access, 29
roof inspections, 121
roof penetrations, 37, 92–93, *93*, 112
roofs
    measurements, 35–37, *35, 36*
    parts of, 9–12, *10, 11, 12*
    role of, 1
    safety, 31–32
root barriers
    about, 7–8, *7, 64*
    installation, 102, 112
    multi-purpose materials, 94
    role, 71
    specifications, 71–72
Ryerson Urban Farm, Toronto, Ontario, 130–131

S

safety, 31–32
saturated loads, 16
saturated weight load test, 83
scaffolding, 29
scuppers, 10, *11*
Sedum
    growth habit, 46
    maintenance, 120
    mats, 55, 56, *57–58*, 88
    suitability of, 44–45
seeding, 55, *57–58*
self-seeding plants, 45, 46
semi-intensive systems
    about, 2–3
    characteristics of, *3*
    weight loads, *16*, 18
shed roofs, 9, 65, 86
single-ply synthetic rubber membranes, 68
site factors
    climate, 38–39
    growing environment, 43–44
    sun exposure, 37
    wind, 39–40
site plan, *41*
site risks, 32
skills, 32
slope
    considerations, *24*, 25–26
    determining, 23–25, *24*
    drainage and, 72–73
    flat roofs, 16, 107–113
    restraint systems, 9, 87–90, *89*
Snell, Clarke, 103
soil, 82–83
    See also growing media
sphagnum peat, 81–82
spring maintenance, 121
standards, 14–15
steel, 94
structural loading capacity, 15–16

styrene-butadiene-styrene (SBS) modified bitumen, 68
succulents, 44–45
sun exposure, 37, 46
surface erosion, 87, 94
synthetic sheet drainage layers, 74

T
temperature, 38–39
tool operation, 32–33
Toronto, Ontario, codes, 15, 94
TPO single-ply membranes, 68
tradespeople, 33
turf roofs, 53

V
vegetation-free zones, 94, 111
vent penetrations, 37, 92–93, 93, 112

W
warm roof assemblies, 86
watering. *See* irrigation
waterproofing assemblies, 64–65
waterproof membranes
    about, 7, 7, *64*
    applications, *70*
    installation, 65–67, 99–100
    role, 64
    specifications, 67–68, 71
water retention layer, 9, 87
weather, installation and, 32, 112
weed cloth, 72
weed control, 117–119, 125
weight loads
    about, 15–16
    allowances, *16*
    example calculation, *17, 18*
    food production roofs, 123
    plants, 58
wind
    erosion protection, 9, 26, 87–91
    site factors, 39–40
window access, 29
winterizing, 121

About the Author

LESLIE DOYLE is the owner of Restoration Gardens, Inc., a green roof design and installation firm that has designed and built dozens of green roofs for residential and commercial buildings from 80 square feet to 30,000 square feet. She is a certified LiveRoof Installer and an accredited Green Roof Professional by the Industry Association, Green Roofs for Healthy Cities. She shares her knowledge and passion for integrating nature in the built environment as a green roof instructor at the Endeavour Centre and formerly at Fleming College. Leslie lives in Toronto, Ontario.

ABOUT NEW SOCIETY PUBLISHERS

New Society Publishers is an activist, solutions-oriented publisher focused on publishing books to build a more just and sustainable future. Our books offer tips, tools, and insights from leading experts in a wide range of areas.

We're proud to hold to the highest environmental and social standards of any publisher in North America. When you buy New Society books, you are part of the solution!

At New Society Publishers, we care deeply about *what* we publish—but also about *how* we do business.

- All our books are printed on 100% post-consumer recycled paper, processed chlorine-free, with low-VOC vegetable-based inks (since 2002). We print all our books in North America (never overseas)
- Our corporate structure is an innovative employee shareholder agreement, so we're one-third employee-owned (since 2015)
- We've created a Statement of Ethics (2021). The intent of this Statement is to act as a framework to guide our actions and facilitate feedback for continuous improvement of our work
- We're carbon-neutral (since 2006)
- We're certified as a B Corporation (since 2016)
- We're Signatories to the UN's Sustainable Development Goals (SDG) Publishers Compact (2020–2030, the Decade of Action)

To download our full catalog, sign up for our quarterly newsletter, and to learn more about New Society Publishers, please visit newsociety.com

ENVIRONMENTAL BENEFITS STATEMENT

New Society Publishers saved the following resources by printing the pages of this book on chlorine free paper made with 100% post-consumer waste.

TREES	WATER	ENERGY	SOLID WASTE	GREENHOUSE GASES
66	5,300	28	220	28,700
FULLY GROWN	GALLONS	MILLION BTUs	POUNDS	POUNDS

Environmental impact estimates were made using the Environmental Paper Network Paper Calculator 4.0. For more information visit www.papercalculator.org